AF389370

Monitoring of Honey Bee Colony Losses

Monitoring of Honey Bee Colony Losses

Editor

Aleš Gregorc

MDPI • Basel • Beijing • Wuhan • Barcelona • Belgrade • Manchester • Tokyo • Cluj • Tianjin

Editor
Aleš Gregorc
University of Maribor
Slovenia

Editorial Office
MDPI
St. Alban-Anlage 66
4052 Basel, Switzerland

This is a reprint of articles from the Special Issue published online in the open access journal *Diversity* (ISSN 1424-2818) (available at: https://www.mdpi.com/journal/diversity/special_issues/monitoring_honey_losses).

For citation purposes, cite each article independently as indicated on the article page online and as indicated below:

LastName, A.A.; LastName, B.B.; LastName, C.C. Article Title. *Journal Name* **Year**, *Volume Number*, Page Range.

ISBN 978-3-0365-3441-1 (Hbk)
ISBN 978-3-0365-3442-8 (PDF)

Cover image courtesy of Aleš Gregorc.

Contents

About the Editor

Aleš Gregorc graduated from the Veterinary Faculty, University Ljubljana, Slovenia. He obtained a Ph.D. in Veterinary Sciences from the same university. He is now a Professor of Apiculture at the University of Maribor, Faculty of Agriculture and Life Science. He previously worked at the Mississippi State Uni., Coastal Res. and Extension Center and was also previously employed by the Agricultural Institute of Slovenia. He studies the effects of pesticides and pathogens at the cellular and molecular level to demonstrate and localize cell death in tissues. He has also extensively worked on developing varroa and small hive beetle control strategies using a variety of experimental protocols in the laboratory and in the field. He was the head of the breeding organization for Carniolan honey bees in Slovenia. He has been active in research and has participated in national and international collaboration projects. He has acted as a PhD supervisor and external examiner for PhDs at several universities. He publishes reviewed scientific articles and participates in national and international conferences, workshops, and professional meetings.

Preface to "Monitoring of Honey Bee Colony Losses"

'Monitoring of Honey Bee Colony Losses' is a research field supported by the COLOSS organization (Prevention of honey bee COlony LOSSes, http://coloss.org/). The book is dedicated to studies and improvements in the area of managed honey colonies' survival and health.

Aleš Gregorc
Editor

Editorial

Monitoring of Honey Bee Colony Losses: A Special Issue

Aleš Gregorc

Faculty of Agriculture and Life Sciences, University of Maribor, Pivola 10, 2311 Hoče, Slovenia;
ales.gregorc@um.si

Received: 16 October 2020; Accepted: 16 October 2020; Published: 17 October 2020

Abstract: In recent decades, independent national and international research programs have revealed possible reasons for the death of managed honey bee colonies worldwide. Such losses are not due to a single factor, but instead are due to highly complex interactions between various internal and external influences, including pests, pathogens, honey bee stock diversity, and environmental change. Reduced honey bee vitality and nutrition, exposure to agrochemicals, and quality of colony management contribute to reduced colony survival in beekeeping operations. Our Special Issue (SI) on "Monitoring of Honey Bee Colony Losses" aims to address specific challenges facing honey bee researchers and beekeepers. This SI includes four reviews, with one being a meta-analysis that identifies gaps in the current and future directions for research into honey bee colonies mortalities. Other review articles include studies regarding the impact of numerous factors on honey bee mortality, including external abiotic factors (e.g., winter conditions and colony management) as well as biotic factors such as attacks by *Vespa velutina* and *Varroa destructor*.

Keywords: honey bee diseases; stressors; pathology; honey bee mortalities; colonies management

---

High mortality in honey bee colonies has been reported worldwide in recent decades without definitive identification of the causes [1]. Several hypotheses have been postulated to explain these losses, but the causes have not been clearly identified [2]. Many factors, including internal and external pressures, exposure to various pathogens, lack of diversity of food sources, management problems, exposure to agrochemicals and a variety of stressors [1,3,4], act in isolation or, more often, in combination, to drive increased mortality among individual bees or managed honey bee colonies [4–6].

This SI includes 6 research articles, 4 reviews, and 13 other pertinent papers addressing honey bees as individual or social organisms responding to a variety of pathogens and nonpathogenic factors such as environmental stressors, honey bee colony management, and beekeeping practices. Many previous studies relied on meta-analyses to interpret the underlying causes of global bee decline, identify gaps in current research, and propose new priorities for research [7]. Authors in this review [7] analyzed 293 international scientific papers. They examined the methodologies used to link various biotic and abiotic stressors to global losses in managed populations of *Apis mellifera*. They concluded an urgent need for standardized testing of the lethality of stressors. The stressors and associated mortality rates in managed honey bee colonies vary globally. Neov et al. [8] describes five stressors responsible for most global declines in managed honey bee colonies. The first stressor is the human-driven spread of pathogenic and pest organisms (e.g., *Varroa destructor* and *Aethina tumida*). More recently, a highly invasive insect predator, *Vespa velutina*, or the Asian yellow-legged hornet have represented a serious new threat to managed honey bees and native pollinators, thus necessitating monitoring and appropriate management actions to prevent further beekeeping losses in affected areas [9]. Varroa (*Varroa destructor*) is the single most significant cause of lower winter survival in honey bees, and improved control of this parasitic mite is still urgently needed [10]. Gregorc and Sampson [11] showed greater effectiveness of Varroa control and greater colony survival with improved diagnostic methods, organic acid treatments and other integrated methods. Moreover, treatment timing is critical for varroa control and improved colony health. For example, control was improved by following-up short-term evaporation of formic acid in summer with oxalic acid trickling in winter. Similarly, in Croatian bees hives, Tlak Gajger et al. [12] used repeated hive monitoring followed up with applications of commercial products to continuously induce mite mortality and prevent reinfestations.

The second stressor involves landscape changes, both positive and negative. Negative changes more typically involve habitat loss or degredation. However, Buchori et. al. [13] identified positive habitat effects while studying the impact of land use patterns, agricultural intensification and insecticides residues on honey bees (*Apis cerana* and *A. mellifera*) and stingless bees (*Tetragonula laeviceps*) in Indonesia (Bogor and Malang regions). They discovered certain habitat types alleviate bee stress and promote colony growth and queen reproductive output, especially forests with abundant sources of honeydew.

The third stressor, intensification of agricultural production, includes the use of fertilizers and pesticides as well as other chemical compounds originating within highly managed colonies or stored honey products. Pesticides even at sub-lethal doses can harm honey bee health and colony productivity. Currently, few studies are addressing the sub-lethal effects of biotic [14] and abiotic stressors [15] on bee health and the vulnerabilities among subspecies of honey bees. Sublethal pesticide effects are not as well studied as lethal effects. There are several potential ways that honeybees can be exposed to sublethal levels of pesticides and other environmental pollutants, for example, through water collection, by contact with foliage, or through contaminated pollen and nectar [16]. The application and proper dosage of pesticides (acaricides) inside the hive to control parasitic mites could also affect the health of brood, worker bees [17], and queens [18]. Therefore, it is of great interest to study the effects of acaricides and other pesticides on honey bee health and productivity. Martinello et al. [19] extensively surveyed, in Italy from 2015 and 2019, the occurrence of pesticide residues in the bodies of dead honey bees, in samples of comb, in bee bread, and on plant tissues (leaves, corn seedlings, and maize). From 696 samples, honey bees were exposed to 150 pesticides, with 50% of the honeybee samples testing positive for one or more active ingredients with an average of 2 and a maximum of 7 pesticides per sample. By analyzing dead bees and plant materials from the field, these studies contributed to a better understanding of the influence of individual or combined pesticide mixtures on honeybee health, even when chemicals occurred at sublethal concentrations. Later, an integrative protocol was developed for monitoring the effects of field-realistic exposure of honey bees to neonicotinoids by monitoring honeybee colony activity along with electronic measurements of internal and external hive temperature and humidity as well as colony weight. It was found that quality samples preparation and follow up of honeybee colonies and honey/pollen flows can be successfully performed by using classic methods to monitor weather conditions; activities and population of colonies; weight gain; or contemporary technologies, including electronic sensors combined with the Internet of Things and big data storage [20].

Pesticides are not the only harmful environmental compounds encountered by honey bees. Managed colonies are vulnerable to a honey-breakdown product, hydroxymethylfurfural (HMF). Gregorc et al. [21] showed that HMF at higher concentrations reduces longevity and midgut integrity of caged worker *Apis mellifera carnica*. Negative effects of HMF on bees after 15 feeding days included extended midgut cell death and increased worker mortality.

The remaining two stressors are attracting increased attention, despite the logistical challenges they pose to researchers. These stressors are related to ecological change brought about by climate change and resulting weather intensification and recent invasions of new, non-native plant species [22]. Moreover, additional studies are also needed that can account for genetic variation (e.g., subspecific differences), which can be profound, in bee responses to these and other stress factors [23]. A citizen science survey identified additional causes of high overwintering losses (15.2%) in Austrian honey bee colonies (2018/19) [10]. These causes were related to certain beekeeping practices that create queen problems (reduced fecundity, reduced lifespan) during the season. Colonies can be affected by introduction pesticides via contaminated food sources, wax from outside a given operation, collecting melezitose, and foraging for a late catch crop.

In conclusion, this SI highlights the most important research or extension topics associated with honey bee colonies mortalities worldwide. I hope readers have gained new knowledge and directions for further scientific work.

Funding: The work was supported by ARRS, Research Program P1-0164 (Research for improvement of safe food and health).

Acknowledgments: I would like to thank Blair Sampson for initial proofreading of the manuscript and staff members at the MDPI editorial office for their support.

Conflicts of Interest: The author declares no conflict of interest.

References

1. Benaets, K.; Van Geystelen, A.; Cardoen, D.; De Smet, L.; de Graaf, D.C.; Schoofs, L.; Larmuseau, M.H.D.; Brettell, L.E.; Martin, S.J.; Wenseleers, T. Covert deformed wing virus infections have long-term deleterious effects on honeybee foraging and survival. *Proc. R. Soc. B Biol. Sci.* **2017**, *284*, 20162149. [CrossRef]
2. Decourtye, A.; Devillers, J. Ecotoxicity of neonicotinoid insecticides to bees. In *Insect Nicotinic Acetylcholine Receptors. Advances in Experimental Medicine and Biology*; Thany, S.H., Ed.; Springer: New York, NY, USA, 2010; pp. 85–95, ISBN 978-1-4419-6445-8.
3. Goulson, D.; Nicholls, E.; Botías, C.; Rotheray, E.L. Bee declines driven by combined stress from parasites, pesticides, and lack of flowers. *Science* **2015**, *347*, 1255957. [CrossRef] [PubMed]
4. Christen, V.; Bachofer, S.; Fent, K. Binary mixtures of neonicotinoids show different transcriptional changes than single neonicotinoids in honeybees (*Apis mellifera*). *Environ. Pollut.* **2017**, *220*, 1264–1270. [CrossRef]
5. Alburaki, M.; Cheaib, B.; Quesnel, L.; Mercier, P.-L.; Chagnon, M.; Derome, N. Performance of honeybee colonies located in neonicotinoid-treated and untreated cornfields in Quebec. *J. Appl. Entomol.* **2017**, *141*, 112–121. [CrossRef]
6. Fine, J.D.; Cox-Foster, D.L.; Mullin, C.A. An inert pesticide adjuvant synergizes viral pathogenicity and mortality in honey bee larvae. *Sci. Rep.* **2017**, *7*, 40499. [CrossRef] [PubMed]
7. Havard, T.; Laurent, M.; Chauzat, M.-P. Impact of stressors on honey bees (*Apis mellifera*; Hymenoptera: Apidae): Some guidance for research emerge from a meta-analysis. *Diversity* **2020**, *12*, 7. [CrossRef]
8. Neov, B.; Georgieva, A.; Shumkova, R.; Radoslavov, G.; Hristov, P. Biotic and abiotic factors associated with colonies mortalities of managed honey bee (*Apis mellifera*). *Diversity* **2019**, *11*, 237. [CrossRef]
9. Laurino, D.; Lioy, S.; Carisio, L.; Manino, A.; Porporato, M. Vespa velutina: An alien driver of honey bee colony losses. *Diversity* **2020**, *12*, 5. [CrossRef]
10. Oberreiter, H.; Brodschneider, R. Austrian COLOSS survey of honey bee colony winter losses 2018/19 and analysis of hive management practices. *Diversity* **2020**, *12*, 99. [CrossRef]
11. Gregorc, A.; Sampson, B. Diagnosis of Varroa Mite (*Varroa destructor*) and sustainable control in honey bee (*Apis mellifera*) colonies—A review. *Diversity* **2019**, *11*, 243. [CrossRef]
12. Tlak Gajger, I.; Svečnjak, L.; Bubalo, D.; Žorat, T. Control of *Varroa destructor* mite infestations at experimental apiaries situated in Croatia. *Diversity* **2020**, *12*, 12. [CrossRef]
13. Buchori, D.; Rizali, A.; Priawandiputra, W.; Sartiami, D.; Johannis, M. Population growth and insecticide residues of honey bees in tropical agricultural landscapes. *Diversity* **2020**, *12*, 1. [CrossRef]
14. Jacques, A.; Laurent, M.; EPILOBEE Consortium; Ribière-Chabert, M.; Saussac, M.; Bougeard, S.; Budge, G.E.; Hendrikx, P.; Chauzat, M.-P. A pan-European epidemiological study reveals honey bee colony survival depends on beekeeper education and disease control. *PLoS ONE* **2017**, *12*, e0172591. [CrossRef]
15. Benuszak, J.; Laurent, M.; Chauzat, M.-P. The exposure of honey bees (*Apis mellifera*; Hymenoptera: Apidae) to pesticides: Room for improvement in research. *Sci. Total Environ.* **2017**, *587–588*, 423–438. [CrossRef] [PubMed]
16. Papaefthimiou, C.; Pavlidou, V.; Gregorc, A.; Theophilidis, G. The action of 2,4-Dichlorophenoxyacetic acid on the isolated heart of insect and amphibia. *Environ. Toxicol. Pharmacol.* **2002**, *11*, 127–140. [CrossRef]
17. Gregorc, A.; Pogacnik, A.; Bowen, I.D. Cell death in honeybee (*Apis mellifera*) larvae treated with oxalic or formic acid. *Apidologie* **2004**, *35*, 453–460. [CrossRef]

18. Pettis, J.S.; Collins, A.M.; Wilbanks, R.; Feldlaufer, M.F. Effects of coumaphos on queen rearing in the honey bee, *Apis mellifera*. *Apidologie* **2004**, *35*, 605–610. [CrossRef]
19. Martinello, M.; Manzinello, C.; Borin, A.; Avram, L.E.; Dainese, N.; Giuliato, I.; Gallina, A.; Mutinelli, F. A survey from 2015 to 2019 to investigate the occurrence of pesticide residues in dead honeybees and other matrices related to honeybee mortality incidents in Italy. *Diversity* **2020**, *12*, 15. [CrossRef]
20. Căuia, E.; Siceanu, A.; Vișan, G.O.; Căuia, D.; Colța, T.; Spulber, R.A. Monitoring the field-realistic exposure of honeybee colonies to neonicotinoids by an integrative approach: A case study in Romania. *Diversity* **2020**, *12*, 24. [CrossRef]
21. Gregorc, A.; Jurišić, S.; Sampson, B. Hydroxymethylfurfural affects caged honey bees (*Apis mellifera* carnica). *Diversity* **2020**, *12*, 18. [CrossRef]
22. Switanek, M.; Crailsheim, K.; Truhetz, H.; Brodschneider, R. Modelling seasonal effects of temperature and precipitation on honey bee winter mortality in a temperate climate. *Sci. Total Environ.* **2017**, *579*, 1581–1587. [CrossRef]
23. Suchail, S.; Guez, D.; Belzunces, L.P. Characteristics of imidacloprid toxicity in two *Apis mellifera* subspecies. *Environ. Toxicol. Chem.* **2000**, *19*, 1901–1905. [CrossRef]

Publisher's Note: MDPI stays neutral with regard to jurisdictional claims in published maps and institutional affiliations.

diversity

Article

IoT-Driven Workflows for Risk Management and Control of Beehives

Charbel Kady [1,2], Anna Maria Chedid [3], Ingred Kortbawi [3], Charles Yaacoub [3], Adib Akl [3], Nicolas Daclin [1], François Trousset [1], François Pfister [4] and Gregory Zacharewicz [1,*]

1. IMT-Mines Ales, LSR, 6 Avenue de Clavières, 30100 Alès, France; charbel.kady@mines-ales.fr (C.K.); nicolas.daclin@mines-ales.fr (N.D.); francois.trousset@mines-ales.fr (F.T.)
2. School of Engineering, Lebanese American University (LAU), P.O. Box 36, Byblos 1401, Lebanon
3. School of Engineering, Holy Spirit University of Kaslik (USEK), P.O. Box 446, Jounieh 1200, Lebanon; annamaria.c.chedid@net.usek.edu.lb (A.M.C.); ingred.g.kortbawi@net.usek.edu.lb (I.K.); charlesyaacoub@usek.edu.lb (C.Y.); adibakl@usek.edu.lb (A.A.)
4. Connecthive, 100 Route de Nîmes, 30132 Caissargues, France; pfister@connecthive.com
* Correspondence: gregory.zacharewicz@mines-ales.fr; Tel.: +33-4-6678-5000

Abstract: The internet of things (IoT) and Industry 4.0 technologies are becoming widely used in the field of apiculture to enhance honey production and reduce colony losses using connected scales combined with additional data, such as relative humidity and internal temperature. This paper exploits beehive weight measurements and builds appropriate business rules using two instruments. The first is an IoT fixed scale installed on one hive, taking rich continuous measurements, and used as a reference. The second is a portable nomad scale communicating with a smartphone and used for the remaining hives. A key contribution will be the run and triggering of a business process model based on apicultural business rules learned from experience and system observed events. Later, the evolution of the weight of each individual hive, obtained by either measurement or inference, will be associated with a graphical workflow diagram expressed with the business process model and notation (BPMN) language, and will trigger events that inform beekeepers to initiate relevant action. Finally, the BPMN processes will be transformed into executable models for model driven decision support. This contribution improves amateur and professional user-experience for honeybee keeping and opens the door for interoperability between the suggested model and other available simulations (weather, humidity, bee colony behavior, etc.).

Keywords: beekeeping; BPMN; hives monitoring; IoT; modeling & simulation; interoperability; sensors; honeybee behavior; Industry 4.0; workflow

Citation: Kady, C.; Chedid, A.M.; Kortbawi, I.; Yaacoub, C.; Akl, A.; Daclin, N.; Trousset, F.; Pfister, F.; Zacharewicz, G. IoT-Driven Workflows for Risk Management and Control of Beehives. *Diversity* **2021**, *13*, 296. https://doi.org/10.3390/d13070296

Academic Editor: Luc Legal

Received: 19 May 2021
Accepted: 23 June 2021
Published: 29 June 2021

1. Introduction

Beekeeping is a branch of agriculture that involves the breeding of bees (*Apis mellifera*) to exploit the products of the hive, mainly honey. *Apis mellifera* is a semi-domestic species that has been widely used since antiquity, not only to produce honey, wax, and pollen, but also for its prominent role in pollination, especially of vegetable and fruit crops [1,2].

According to the European Professional Beekeepers Association [3], many statistics that pertain to beekeeping consider professional beekeepers as only full-timers. However, in common practice, we can mainly distinguish three categories of beekeepers according to the importance of their operation. The first, amateur beekeepers, who have a family-friendly practice, own up to 50 beehives. The second category is made up of professional beekeepers who can own more than 150 beehives. Finally, beekeepers who own between 50 and 150 hives are considered as semi-professionals. Generally speaking, professional and semi-professional apiculturists reallocate their hives with respect to the blooms whereas amateurs do not necessarily.

In his 2009 book, Nicola Bradbear [4] stated that pollination is an indirect use value that stems from apiculture. In fact, insects, either directly or indirectly, are responsible of around one third of the pollination process of all plants, or plants products, consumed by humans. Among all pollinating insects, bees play the major role. In Western Europe, for instance, the worth of bee pollination is assessed to be 30–50 times the value of direct honeybees' production [5]. In Africa, bee pollination is sometimes estimated at 100 times the value of direct production. Thus, it is estimated that the economic value provided by bees during pollination is out of all proportion to that of bee products.

As a response to colony collapse disorder (CCD), where major decrease in honeybee colonies is reported all over the word [6–8], advancement in technology is playing a major role in preserving honeybee's species, pollination, and biodiversity sustainability [9,10]. In that context, Industry 4.0 and IoT technologies are playing a major role [11], where the use of sensors, networking, and artificial intelligence (AI) in addition to models, simulations, and process discovery are transforming current industrial processes [12–14]. In fact, the implementation of technology in Beekeeping dates to 1907 when Gates [15] measured temperature for many consecutives days.

With technology and IoT evolving, more accurate, small size, and low-price sensors became available in the market, ensuring a non-disturbing data collection environment [16,17]. Thus, more measurements related to the honeybee colony can now be collected, such as image [18], relative humidity [19], sounds, internal temperature, weight, forager traffic, and gases [20,21].

In 1990, Buchman and Thoenes [22] introduced continuous weight monitoring using a precision scale connected to a computer and collected data every 15 min for one month. Subsequently, many other studies showed a correlation between weight fluctuation and activities occurring inside the hives [23]. For instance, in 2008, according to Meikle et al. [24], the hourly weight, measured using electronic scales linked to data loggers, was calculated to derive the weekly running average weights. Later, the weights associated with the changes in food stores were correlated with humidity ratio changes and foraging activities, which were classified as daily fluctuations. In 2013, Human et al. [25] stated that the weight of full colonies can be measured to examine the nectar flow occurrence during the foraging season or daily gain in nectar stores, the decline of food stores during nonforaging periods [26], and the occurrence of swarming events [24]. In addition, the authors provided examples of applications of honey hives weight measurements networks in different countries, such as Germany, USA, Denmark, and Switzerland. In 2015, Sandra Kordić Evans [27] proposed an embedded hive system that measures weight, temperature, and movement using an accelerometer. Meikle et al. [28], in 2016, added to their previously mentioned work with a temperature analysis by installing sensors in different positions in the beehive and found the corresponding position by correlating each position's temperature variations with its exterior conditions. Anand [29], in 2018, proposed a system that uses many combinations of the weight with sounds, relative humidity, and brood temperature to detect honeybee swarms based on several techniques and methods.

The present study does not yet contain complete results. The described items are examples of apiary data coming from an existing shared database [30] (Section 2.2.1) with other measurements provided by *Connecthive* [31]. Starting from such items, we propose a methodology based on pattern recognition within data recorded through both a static hive scale on one hand and a portable nomadic hive scale on the other hand. The proposed method focuses on weight measurements, and will be considered by introducing two types of measurement: an on-board IoT with richly measurements installed on one witness hive and a nomadic weight measurement scale system used for the adjacent hives. Later, the data collected will be assessed and combined on the server (Sections 2.2.2 and 2.2.3). Finally, this method will contribute to reduce the overall cost by proposing the nomad scale solution instead of installing a scale for each hive.

Moreover, the back-end server will be responsible for the processing and labeling of the collected raw data by discovering recurring patterns associated with events occurring inside the hive. Then, detected patterns will automatically trigger a series of predefined actions based on workflow rules and static models built with the help of experts in the domain (the community of beekeepers using the system) (Section 2.3.2). Finally, all the above will be orchestrated within a user-friendly interface on a smartphone or a tablet where the beekeeper can easily monitor his colony, perform his daily tasks, respond to alerts for possible malfunctions, and forecast his future needs for supplies. As beekeeping tasks cannot be automated, and personal intervention is mandatory, the proposed system will help the beekeeper to plan several relevant tasks precisely: raising and breeding new colonies, feeding the colonies, adding supers, planning sanitary operations, controlling varroa mites, planning operations such as hibernation, etc.

As the system will not be dedicated only for professionals, but also for amateurs, it will encourage a larger number of people to be engaged in the beekeeping field, which can have a positive impact on the environment, biodiversity, and society. To achieve that purpose, a gamification approach will be followed by building a user-friendly interface with voice recognition and image capture features that facilitate the management of a large numbers of hives.

Gamification as a term was introduced for the first time in 2002 by Nick Pelling but it did not gain people's attention until year 2010 [32]. Pelling originally defined gamification as "transferring game environment user interface to the electronic world by making it an entertaining and fast user experience". Nowadays, gamification can refer to "implement main game design features into a real-world application" [33] to motivate and attract users with a professional application in an engaging game-like context [34].

In this context, the system will propose step by step actions and countermeasures to be taken with respect to a set of notifications and alerts reflecting the beehive status in real time. Finally, precision beekeeping [35] is a method of beehive management that relies on the monitoring of honeybee colonies for the reason to minimize supplies expenditure and maximize the honey yield production. Just like the principles already existing in precision agriculture, it is divided into three phases: data collection, data analysis, and the application of decision support for the beekeeper. While collecting the data, measurements are gathered directly from the hives: weight, temperature, and internal humidity, etc. The data analysis phase makes inferences from predefined models often based on artificial intelligence or on expert systems. During the decision support phase, recommendations are submitted to improve the performance of the apiary. Thus, beekeeping, like other fields of agriculture, must initiate a digital revolution. As the conditions of beehive exploitation becoming more and more complex, particularly due to environmental factors, the management of the farms must be rationalized to remain profitable, to produce better quality honey, and to minimize the losses of livestock [36]. Indeed, it is essential to maintain the population of honeybees without which the pollination of crops could no longer be ensured, causing considerable economic and societal damage [37]. The goal here is to use technologies based on a workflow engine, including business rules, and deep learning to help the beekeeper to better manage his bee colonies by monitoring and automating some of his decisions. In this context, the decision support system makes it possible to predict the evolution of each colony and suggests certain breeding operations to be carried out to improve the productivity and survival of the colonies. The contribution of digital techniques should pave the way for precision beekeeping, to minimize invasive treatments and synthetic inputs [38].

2. Materials and Methods

2.1. Experimental Scenario and Methodology

A decision support system will be installed in two professional beekeeping farms (more than 200 hives), in 4 semi-professional farms (more than 50 beehives) and 4 amateur farms (less than 20 beehives).

Then, measurements carried out continuously are collected on the control hives by means of the static scales as well as the discontinuous measurements carried out by means of the mobile scales.

In addition, data will be collected from beehive visit reports entered by beekeepers using the apiary monitoring application on their smartphones. These reports are equivalent to the ground truth and are used to label the data.

Subsequently, a data mining phase will be intended to find emergences of the processes initiated by beekeepers. Therefore, the identified processes are formalized with BPMN. They are differentiated by the type of beekeeping (professional, semi-professional and amateur) and they are placed in a catalog. Processes are validated (method with expert observation, or through a simulator that will be developed using a multi-agent system) and this process catalog will be available to all beekeepers. The system uses it to issue detailed operational advice.

Figure 1 illustrates the proposed methodology which consists of the three main steps: data collection, build time, and run time. Each step, in addition to the relations between different pillars, will be fully discussed in the upcoming sections.

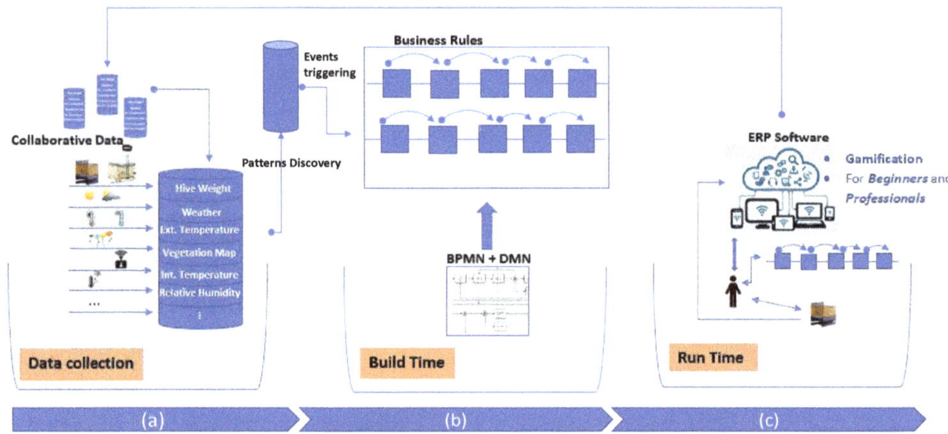

Figure 1. Illustration of the proposed methodology. Our proposed methodology is divided into three main parts: data collection (**a**), build time (**b**), and run time (**c**).

2.2. Data Collection

As mentioned in Section 1, the proposed system is still under development, and although, only the weight measurement collection will be considered in this paper, it is planned to add other measurements as shown in Figure 1a. Data will be collected from different sources such as: hive's weight, weather, external temperature, vegetation and flowers maps, internal temperature, relative humidity, etc. In addition to the mentioned sources, collaborative data will be gathered from different users to enhance system's performance, especially in business rule discovery and simulation.

Whereas many data collection tools will be added to system in the future, the newly proposed weight measurement method will be fully discussed in the following paragraph. This method involves two types of weight measurements, namely continuous and discontinuous, as shown in Figure 2.

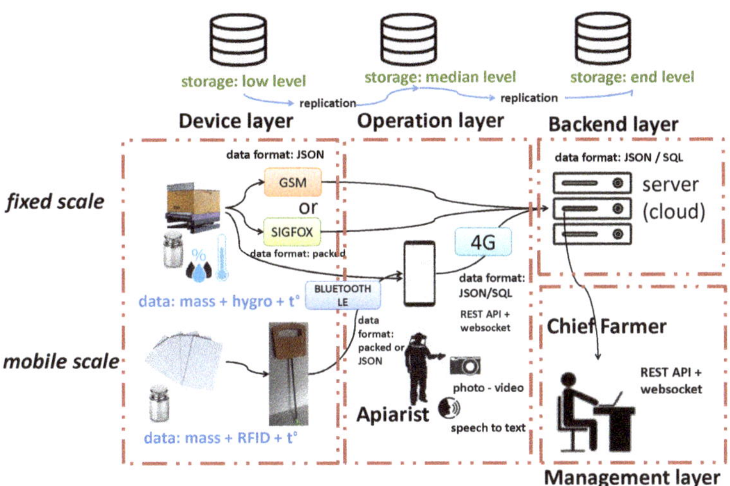

Figure 2. Implementation and Communication between different layers.

2.2.1. Datasets

Our data were obtained from HOBOS public dataset [30] with additional measurements provided by *Connecthive* [31]. HOBOS data originate from Wurzburg and Schwartau in Germany for the period between 1 January 2017, and 19 May 2019, while Connecthive data were collected from Villefranche de Rouergue in France between February and November 2019, as shown in Table 1. This table shows, from left to right, the location, the number of hives, the observation period, and the sampling frequency. The weight of the beehive through time is measured in Kg for the hives of Wurzburg and Schwartau, while that of the hives of Villefranche de Rouergue is measured in grams. Note that, between May 2018 and October 2018, there is missing data from the Wurzburg hive because the system was down. It is important to mention that the temperature of the beehive through time, the humidity of the beehive through time, and the number of departures from and arrivals to the beehive for a certain date are available with HOBOS dataset, but they have not been used in the current study.

Table 1. Weight data measurement conditions.

Location	Number of Hives	Observation Period	Sampling Frequency
Wurzburg	1	2017 to 2019, 1 January 2017 to 19 May 2019	1 sample/min
Schwartau	1	2017 to 2019, 1 January 2017 to 19 May 2019	2 samples/day (at 1 a.m. and 1 p.m.)
Villefranche de Rouergue	20	February 2019 to November 2019	2 samples/h, up to 1 sample/min

2.2.2. Global Collection of Information on Beehives

In the market there exist several companies that provide tools and methods for global data collection from the beehives, as shown in Table 2. We can mention several examples such as: BTS, Arnia [39], ApisProtect [40], Beeehive Scales [41], BuzzTech [42], Solution-Bee [43], etc. Moreover, open data sources also could be available online as shown in Table 1. Although hives measurements performed by Arnia [39], for instance, might look the same, the reason behind the choice of Connecthive [31] over similar service providers is, first, that open-source data exists on the internet as fixed raw data which cannot be customized for the needs of our experiments nor extended, and second, the association of a nomadic scale and a fixed scale shown in Section 2.2.3, in addition to RFID tag and the automatic weight recording in the database for each single hive are unique and specific to

Connecthive, as there is no similar approach by other compagnies. Moreover, the workflow approach, a specific subject of this paper, will be elaborated more in Sections 2.3.2 and 3.3.2.

Table 2. Similar solutions comparison.

	IoT	Humidity	Weight	Sound	Temp.	Theft Alarm	Realtime Data	Alerts	Mobile App.	Nomad	RFID	Workflow	Smart App
Arnia [39]	✔	✔	✔		✔	✔	✔	✔	✔				✔
ApisProtect [40]	✔	✔		✔	✔		✔	✔	✔				✔
BeeHive Scales [43]	✔	✔	✔		✔		✔	✔					
SolutionBee [43]	✔		✔		✔		✔	✔					✔
Broodminder [44]	✔	✔	✔		✔				✔				
Osbeehives [45]	✔	✔		✔	✔	✔	✔	✔	✔				
ConnectHive [31]	✔	✔	✔		✔		✔	✔	✔	✔	✔	✔	✔

Connecthive is a company that develops beekeeping data acquisition systems based on the one hand, on measurements taken from on-board sensors in a few witness hives (so-called richly measured), and on the other hand, from a nomadic acquisition system associated with a voice input interface on tablet or smartphone. The nomadic system is moved to all the hives (said to be weakly measured) during the visits made by the beekeeper. Weight measurements are smoothed out to disregard variations caused by wind. Regarding the support, the hives are placed on a horizontal support even if the ground is not. The data collected are fed back in real time or in deferred mode (thanks to an IoT network) on a server which aggregates it. A rich measurement includes the following quantities in particular: instantaneous weight of the hive, internal temperature, and hygrometry of the swarm, rustling of the swarm, outdoor temperature and hygrometry, wind speed, duration of sunshine, inbound and outbound bee traffic, level pollen collection. A weak measurement includes the following quantities: hive weight, swarm size measured by thermography, instantaneous traffic level capture (short videos), automated varroa count (still image processing). In addition, observed but non-metrological observation data supplement this weak measurement, in particular: surface area and type of brood, number of frames occupied, possible pathological signs.

Although in this study only continuous and discontinuous weight measurements will be considered, the integration of other input data will be part of the future work.

As for the data transmission, GSM, SIGFOX, and 4G are used for communication between the four main layers, i.e., device layer, operation layer, backend layer, and management layer, as described in Figure 3. In the case of network shortage, the replication mechanism synchronizes the same data among the three layers as soon as the respective networks (GSM, SIGFOX, and 4G) are reached. The data structures increase in complexity as the layer is higher. Finally, it is essential to mention that, in Lebanon, there is no SIGFOX. Thus, GSM and 4G are only considered.

2.2.3. Weight Measurements

As illustrated in Figures 2 and 3, continuous measurements are taken by a stationary scale connected to the cloud, whereas the discontinuous measurements are carried out with a mobile (portable) scale, which is also connected to a network. Beehives weighed with the mobile scale are automatically identified with an RFID tag and the weight data are recorded in the user's smartphone before being consolidated in the cloud.

The discretization of measurable data into events associated with a finite time is the foundation of today's computation theory [46]. Data from observed events are stored after appending it with a timestamp. This discretization is due to the clock-based nature of computers. However, the physical time is continuous, and physical quantities not acquired by the sensors (e.g., physical quantities at time instants different from sampling times) can still be estimated by interpolation or extrapolation, thus making the collected data virtually continuous. Signal processing techniques usually consider the continuous nature of such measurements and lean towards representing signals as mathematical functions. Meanwhile, on the other hand, discrete processing of large amounts of data might require enormous storage resources, e.g., in case of a large number of sensors, and/or data observation spanning a long period of time.

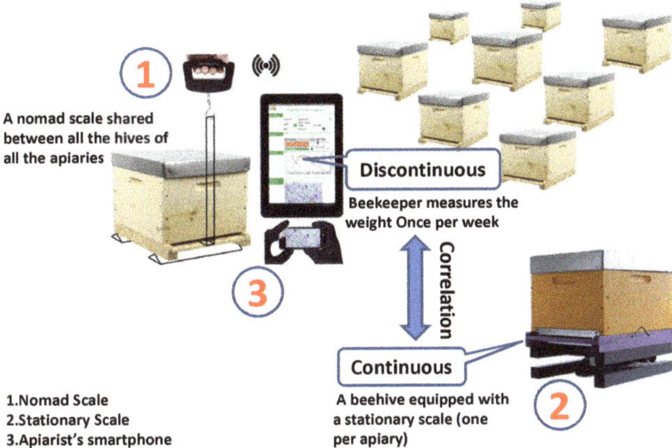

Figure 3. Continuous and discontinuous weight measurements.

Discrete measurements can be considered continuous in case the sampling interval is as small as the sensor's time resolution. For instance, if a sensor is capable of providing a maximum of one sample per second, then acquiring a measurement every second is continuous relative to that sensor's performance. In this project, we exploit this type of continuous measurement, namely instantaneous weight measurements of a reference hive, coupled with discrete measurement samples, approximately a week apart, acquired from all other (non-reference) hives in the apiary.

As mentioned earlier, the proposed system relies on external weight measurements. Because the weight measurement of each beehive is too expensive, the weight of one beehive per apiary is continuously measured and the weight of all the beehives is measured one time each week. Therefore, the challenge resides in inferring continuous weight variations of each beehive from the variations of the continuous measures. To do so, two types of scale are used: a continuous (stationary) and a discontinuous (nomad) scale. More precisely, the continuous scale is used to measure different parameters of the hive over a specified region, using a mobile sensor. A fixed device can be associated with this mobile sensor under the control hive which is later equipped with fixed sensors for weight, temperature, and videos analysis. All measurements are stored in the database to use them to control a hive simulator. The constant recording of the weight, temperature, and humidity, applied over the control hive, relates to the spot records made on each hive separately, using the discontinuous scale. These accumulated data can then faithfully reflect the situation of the colony by giving it a unique profile using artificial intelligence (AI).

Note that the continuous scale is connected to the IoT by Sigfox or by mobile data networks (2G–3G). It can also be disconnected. In this case, data are uploaded to the apiarist's smartphone when nearby, using Bluetooth. The discontinuous scale is always connected to the apiarist's smartphone. An illustration of the continuous and discontinuous weight measurements is presented (Figure 2).

2.3. Build Time

We carried out a proof of concept based on the data coming from two apiaries: Wurzburg (Center of Germany) and Villefranche de Rouergue (South of France). The results of this proof of concept are not sufficient as apidological results but allow us to design the methodology of our project. After a simultaneous data collection, weight variation patterns will be identified for data labeling and events association. These events will trigger on the predefined BPMN workflow model a sequence of automated business rules in response to the hive's events. Weight patterns discovery and BPMN workflow will be discussed in Sections 2.2.1 and 2.2.2.

2.3.1. Weight Patterns Detection and Analysis

The study is divided into four main sections: (1) provision of datasets containing the weight and ambient temperature; (2) loading, cleaning, and smoothing the data; (3) analysis of weight and temperature fluctuations; and (4) results validation.

- Data preprocessing

In the first phase, datasheets containing the recorded weight with the corresponding timestamp are cleaned. All invalid date formats and unusual weight values are eliminated. For an enhanced analysis, the smoothing is used to eliminate minor variations in the data and to make the difference between the samples uniform. In addition, the smoothing method calculates the average of the weight between two samples having the difference in time equal to the window given.

- Data analysis

The analysis is divided in two parts: the monthly and daily analysis. In both parts a linear regression is fit to the model used to find a correlation between the time and the weight. The coefficient of determination R2 and the slope are both used to interpret the variations.

In the monthly study, events are analyzed depending on weight variations and the current season. The data are resampled to 12 h and sets of monotonous samples were analyzed consequently. Each set follows a linear regression, and the slope is then used to give the appropriate interpretation for the remarkable changes in a single frame.

In the daily analysis, the parameters considered in the study are the weight, the time of the day, and the outside temperature that is categorized into three sets. The data are resampled to 1 h, then split into three parts: the inactive period, from midnight until dawn and dusk until midnight, and the active period from dawn until dusk. In each part, monotonous segments are fit to a linear regression model, where the slope is utilized to determine the corresponding events. As a first step, three temperature ranges are identified too cold (below 12 °C), normal (between 12 °C and 35 °C), and too hot (over 35 °C) [47]. Then, in each range and depending on the time of the day, an appropriate interpretation is generated. Finally, all detected events and patterns are displayed on daily and monthly plots that show each continuous weight variation analysis in a specific color along with its corresponding label. The results are validated from independent datasets from France and Germany.

The data is smoothed using a running average window. Noise resulting from beekeeper intervention or system failure is excluded from the dataset. Below, Figure 4 illustrates examples of detected patterns based on data collected by Connecthive and correlated with beekeeper observation from the hive.

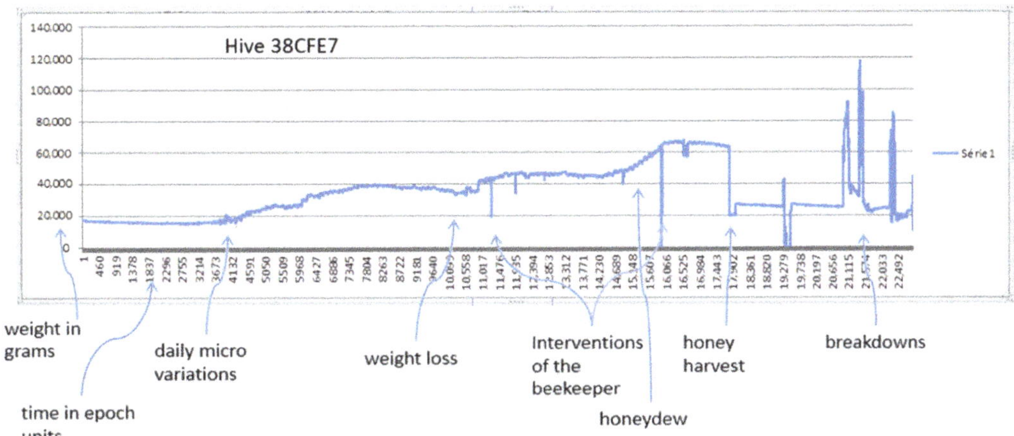

Figure 4. Detected Patterns.

2.3.2. Business Process Model and Notation (BPMN) & Workflows

Regarding apiculture best practices, two business process models are built. The first model is a result of face-to-face interviews with three different amateurs beekeepers in Lebanon. All common practices were validated, and a comprehensive synthesis is built accordingly before building the model itself. The resulting (BPMN) model for amateur beekeepers consists of around 60 gates. As for the second model, it is built with professionals in the domain and is an outcome of many meetings with a leading company in Lebanon, l'Atelier du Miel [48], where to better organize the received information, an excel file was created to describe the beekeeping process taking into consideration several parameters, such as hive's status, region, altitude, etc. In addition, Connecthive [31] also provided us with pseudocode that illustrates some of the beekeeping processes shown in Figure 5. The resulting (BPMN) model for professional beekeepers consists of more than 100 gates.

Table 3 illustrates an excerpt from the mentioned Excel file as the result of many interviews with l'Atelier du Miel [48] correlated with provided data from Connecthive [31], France during spring season and fall respectively. The table describes as mentioned previously: the region, altitude, the starting day, the season, flower types and hive's status. Moreover, in the table, we can distinguish two main categories of tasks with their respective actions and countermeasures: regular daily tasks and anomalies detection. Please note that this is only an excerpt of the original file and the provided information in the table is not exhaustive but is given just as an example. Finally, the same type of anomalous events provided by Connecthive and shown in the bottom of the table, will be fully discussed in Section 3.3.

In this paper and to avoid complexity, only a simple model will be considered taking only into consideration the hive's weight and the season and was built from the pseudo code shown in Figure 5. The BPMN model will be shown later in the results section (Section 3.3.2). To perform the modeling, business process model and notation (BPMN—OMG, 2013) [49] is deployed. BPMN 2.0 makes available a set of modeling objects allowing any organization (industrial, health, military...) to represent their business process.

The aims of BPMN are to be "readily understandable by all business users, from the business analysts that create the initial drafts of the processes, to the technical developers responsible for implementing the technology that will perform those processes, and finally, to the businesspeople who will manage and monitor those processes". At the descriptive layers, the proposed modeling objects allow to build a process model to understand what an organization do and how it works [50].

Table 3. Extract from the beekeeping best practices Excel file.

Atelier du Miel	South, Lebanon				
Altitude	**Starting Day (MM/dd)**	**Season**	**Flower**	**Hive's Status**
(0–450 m)	**02/15**	**Spring**	**Citrus & Avocado trees**	**Colony Developpment**
Regular Tasks		**Actions taken to be taken**		
		Check for colony expansion		
Checkup every three days		Check for supplies		
		Check anomalies		
		Check if eggs are placed from middle to out		
Check queen strength		Check queen activities (Circular, fast, organized)		
. . .		. . .		
Anomalies		**Actions taken to Countermeasure Anomalies occuring in the hive**		
Old queen		Terminate (Could be terminated and replaced)		
		Treat and reallocate for one week to avoid infection		
If Varroa Detected		Keep checking till varroa is cleared		
		Return the hive if varroa free		
		Sperate/Eliminate infected frames from the rest		
If Mold detected		Eliminate any cause of humidity		
. . .		. . .		
Connecthive	**Villefranche, France**				
Altitude	**Starting Day (MM/dd)**	**Season**	**Flower**	**Hive's Status**
(0–450 m)	**09/15**	**Fall**	**Few flowers**	**Wintering Mode**
. . .		. . .		
Anomalies		**Actions taken to Countermeasure Anomalies occuring in the hive**		
Shortage in honey Supply		Feed with Liquid Proteins and check for absorption.		
		In case there is no absorption, feed with Candyboard.		
. . .		. . .		

```
1    #Apiarist simple algorithm
2    Start
3        FeedingDeadline←"August28"
4        USERINPUT HiveID
5        USERINPUT prov
6        if prov≠"ok"
7            Weight←GetWeight(HiveID)
8            if date ≤ "FeedingDeadline"
9                LiquidFeeding()
10               GetWeight(HiveID)
11               if Weight <35
12                   CandyboardFeeding()
13                   GetWeight(HiveID)
14               Endif
15           else
16               CandyboardFeeding()
17               GetWeight(HiveID)
18           Endif
19       Endif
20       Wintering()
21   End
```

Figure 5. Simple Pseudo Code, provided by Connecthive, France.

At the analytic level, BPMN allows to deploy key performance indicators (KPI) to make analysis of processes to improve performance. To this purpose, first, BPMN is easily extendable to integrate specific attributes, concepts, or else relation to a process model and to use it as analysis entities [51]. Second, tools currently available to use BPMN often integrate modules for process analysis and continuous improvement.

Lastly, at the executable level elements, such as data used, actors involved within the process are added to run the process. In the same vein as for the analytic level, tools available to integrate the workflow engine are able to interpret these elements and to execute processes.

Regarding the last point, the concept of workflow is considered. The notion of workflow emerged in the late 1960s [52] in the context of the evolution of information systems [53]. Several definitions have been proposed in the literature for the concept of workflow, including that of the workflow management coalition (WFMC), an organization that defines standards for the development of workflow tools [54,55]. Thus, a workflow consists of automating the flow of information in a process by providing each actor with the information necessary to carry out the activities that make up the process by following procedural rules. This is the reason why the process is modeled first (e.g., using formal models BPMN/DMN presented hereafter) to identify the parts that can be executed by a workflow engine [56,57] in order to drive the taking charge of a protocol.

3. Results

3.1. Weight Patterns Detection and Analysis

Results of the monthly analysis showed that negative slopes were linked to the extraction of honey and the number of bees declining in summer/spring whereas it symbolized the death of bees in winter/fall. Positive slopes showed honey production, adult bees forming and occasionally beekeeper intervention in the summer/spring. Figure 6 shows normal variation over one month in spring, moisture changes (in red), adult bees forming and honey production (in blue), and some loss in the number of bees and honey store (in green).

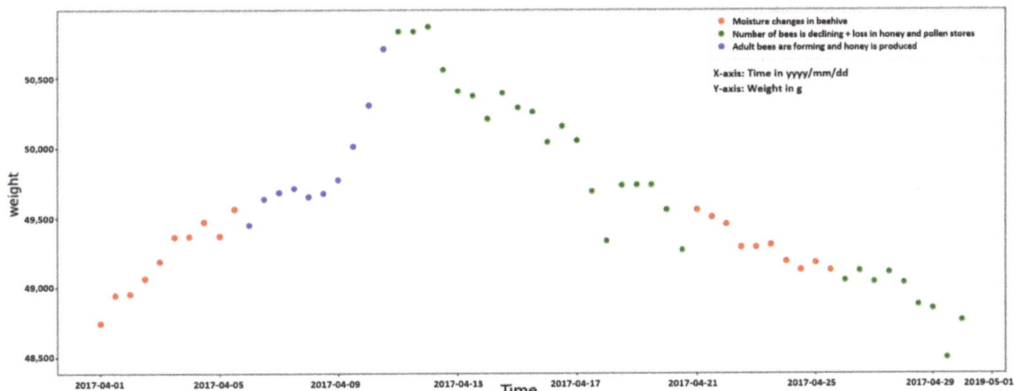

Figure 6. Weight analysis over the month of April showing dryness and food consumption, outgoing bees, and incoming bees with nectar collection during spring season.

Regarding the daily analysis, for temperatures lower than 12 °C, results showed that weight fluctuations were often linked to ventilation, food consumption or beekeeper intervention during active periods, or dryness and moisture during inactive periods since it is considered too cold for the bees to leave the hive. Temperatures between 12 and 35 °C were found to be optimal for foraging activities. In the active period, weight fluctuations were mainly due to bees leaving and entering the hive, whereas dryness and moisture changes affected the weight in the inactive period. As for temperatures higher than 35 °C, the weight fluctuations would mostly be limited to ventilation and beekeeper intervention in active periods, and water vaporization and moisture change during inactive periods. Figure 7 shows a day with a normal temperature where dryness and food consumption (in magenta) occurred in the inactive period, while during the day, bees' foraging activity (in blue) was observed.

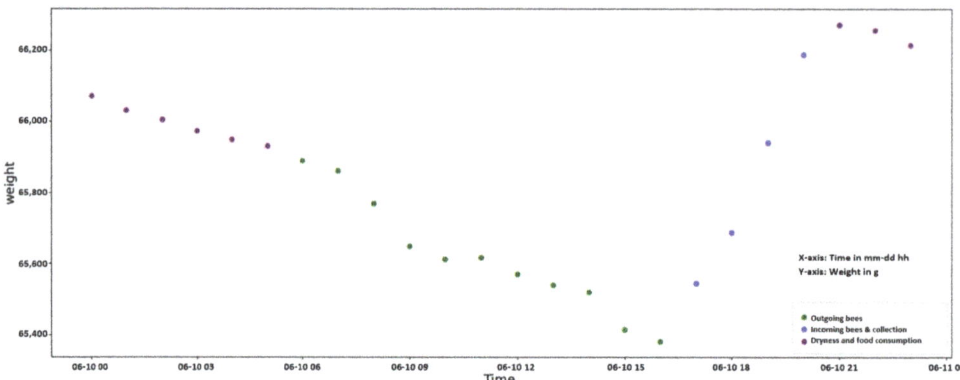

Figure 7. Hourly analysis showing moisture changes in the beehive, adult bees forming and honey production and number of bees declining summer season.

3.2. Artificial Intelligence (AI) and Neural Network

In our datasets, a timestamp is associated with each weight sample. Thus, weight variations of monitored hives can be seen as time series that can be learned by AI models. Such models can be used to complement scarce weight samples obtained from the nomad scale for different hives (c.f. Section 2.2.1), thus inferring continuous weight variations that can be analyzed to identify patterns of anomalies or action-triggering events. Furthermore, AI models can be used to predict future weight variations following an initial observation interval, such that the beekeeper can be alerted about any potential anomaly or trigger ahead of time, allowing for an early intervention whenever necessary. This can also be of utmost importance whenever the monitoring device goes down, due to a power failure or defective sensor for example, thus providing an uninterrupted flow of hive weight data which can reduce the risk of unreliable pattern detection and faulty triggers.

In its early stage of development, a recurrent neural network, namely long short-term memory (LSTM) [58], was designed to predict future weight variations for a beehive. Our model consists of three LSTM layers of 32 units each, separated by a dropout layer with dropout probability of 25%. A sliding window feeds the input layer with a number of consecutive input weight samples to predict the next sample value at the output layer which consists of a single node. The LSTM architecture is shown in Figure 8.

The model was trained and tested with weight measurements from Wurzburg and Schwartau in Germany and Villefranche de Rouergue in France. For each hive, the first 50% of weight samples are used to train the model. Training is performed for 100 epochs with Adam optimizer [59]. Figure 9 highlights a significant weight loss over two months in the late summer-fall be-ginning causing the beekeeper intervention to provide nourishment to the bees, based on Schwartau hive data. After roughly one month of continuous weight loss, the next five-day measurements are fed to the model. Following this initial observation time, the model is used to predict weight variations for the next two weeks. A sliding window considers 10 most recent samples (i.e., five-day peri-od, two samples/day) to predict the expected hive weight within the next 12 h. This window is moved repetitively until a critical weight requiring the beekeeper's intervention is detected. The figure shows that the model was able to predict on 17-08-02 that the hive weight might reach 59 Kg on 17-08-26, with an error of 1 Kg compared to the real measured weight, representing a total loss of 9 Kg within a period of 2 months. The model also predicted a beekeeper's intervention to occur on 17-08-27, one day earlier than the real beekeeper's intervention, as it was trained on data with such beekeeping practice after a significant weight loss within a hive. Without any intervention, further weight loss would be expected for the hive.

Therefore, the model allows the system to trigger an early alarm, alerting the beekeeper of a required intervention ahead of time.

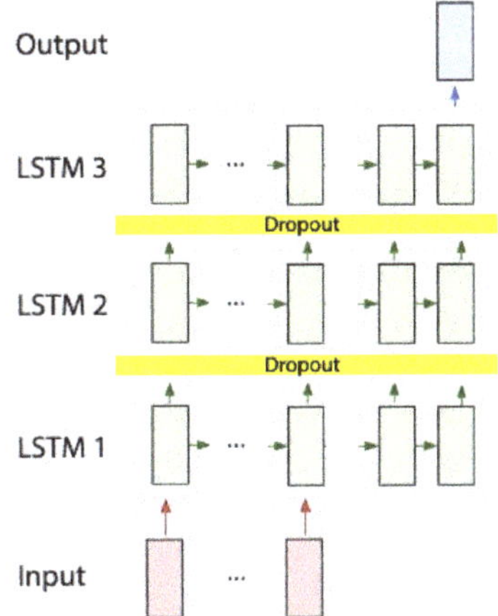

Figure 8. LSTM architecture.

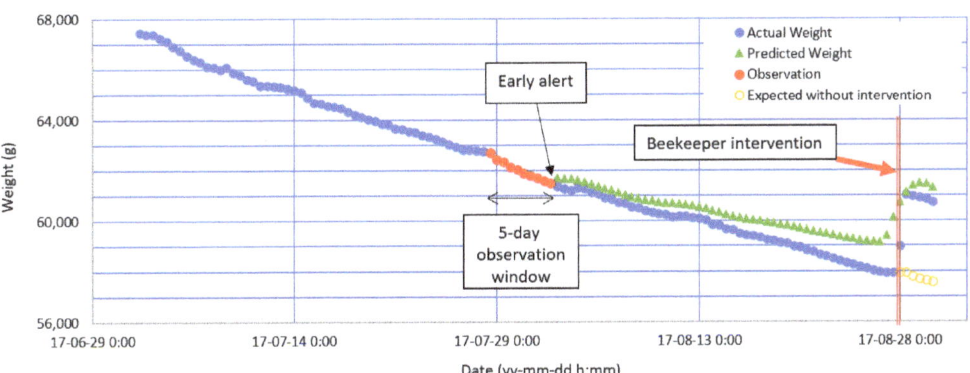

Figure 9. Two months analysis showing significant weight change triggering the beekeeper intervention to provide nourishment in early fall season, with both real and AI-predicted data.

3.3. Process Models and Gamification Approach

3.3.1. Gamification

"Getting things done" (GTD) [60] planning will provide the user with an easy and smooth experience whereby he will receive step-by-step notification on his mobile or tablet concerning where he will be interacting respectively with each move in a simple, gamified, and clear way (Figure 9). In addition to its superfast speed, the application will be designed to generate statistics, work in offline mode, and synchronize across multiple devices.

The system is designed to gain users engagement in the beekeeping domain, regardless of their previous experience in the field, in a consistent, easy, and joyful experience based on a rich data base system (Figure 1a), with a game-like user interface, and a collaborative ecosystem that will be developed later as part of future work.

The novelty of our gamified approach resides in the workflow engine that drives users' actions. The workflow is described in BPMN language and detailed in Section 3.3.2.

3.3.2. BPMN Model

As previously stated in Section 3.3.1, the BPMN is driving the gamification concept by involving the user in the process and guiding the activities to be done. For each event, a series of respective actions will be triggered and pushed on the user's mobile in the form of a notification, in a step-by-step gamified way, similar to the gaming environment, wherein the user can achieve a level only by completing all required steps.

As previously discussed in (Section 2.3.2) for simplicity, a simple BPMN model (Figure 10) was built after the pseudo code (Figure 5) provided by Connecthive [31].

Discovered patterns from the processed data, previously discussed in Section 3.1, will trigger events on the BPMN model and the user will be notified on his mobile phone and respective action will proposed either to perform a daily task or to countermeasure any malfunction occurring in the beehive, as shown in Figures 10 and 11. Moreover, this model will not only be used to trigger events according to the present time data, but it can also be used later in simulation to forecast future anomalies or simply to predict the hive's needs for wax and other supplies.

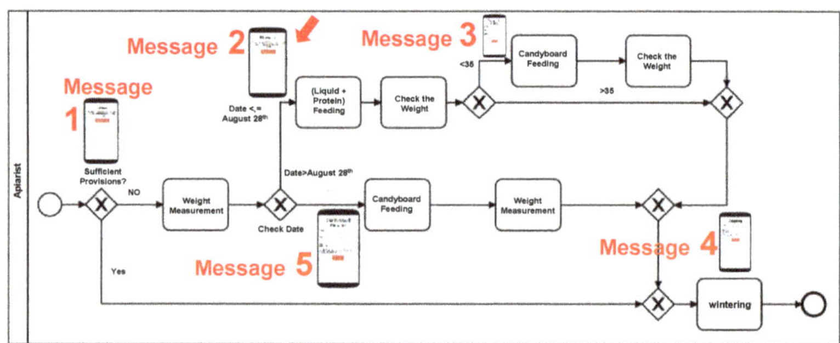

Figure 10. A Simple BPMN model illustrating simple apiarist work.

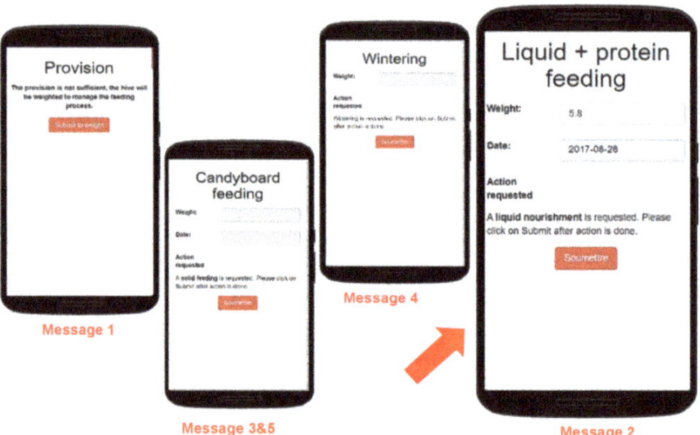

Figure 11. Showing messages 1,2,3,4 and 5.

The red arrow in Figure 10 indicates a critical decrease in the hive weight previously detected and discussed in Section 3.2 as well as shown in Figures 9 and 11. This critical decrease will trigger on the BPMN model an event to feed with liquid nourishment. This event will be communicated to the apiarist in the form of a message displayed on his mobile as shown in Figure 11.

In the above suggested model, an apiarist receives an informative message (1) about the provisions status at the hive, and if sufficient, the system will inform the user to enter the wintering mode (4). If insufficient provisions are detected, the system will enter in weight measurement mode. In this case, the assessment will be based on the date. In case the date is less or equal to 28 August, the apiarist will be alerted in a message (3) to nourish the hive with 6 kg of liquid feeding. After that, the beekeeper will check for the absorption of the nourishment, and if the weight is not reached, the process will be forwarded to candyboard feeding, where the colony is provided at one time with a sufficient quantity to reach 35 kg. The weight will be checked, and the hive will enter the winter mode without waiting for absorption. In contrast, if the date is exceeding 28 August, the apiarist will be alerted in message (5) to feed with solid food (candyboard), check the weight without waiting for absorption, and finally enter wintering mode. Messages 1–5 are illustrated in Figure 11.

To obtain a quick and clear insight about what the output and the messages could look like, Bonita software was used, as illustrated in Figure 12. It has been already deployed on smartphones for test and validation, as presented in Figure 11, on pilot hive farms. Access and response time are already satisfactory in case the cellular network is sufficient.

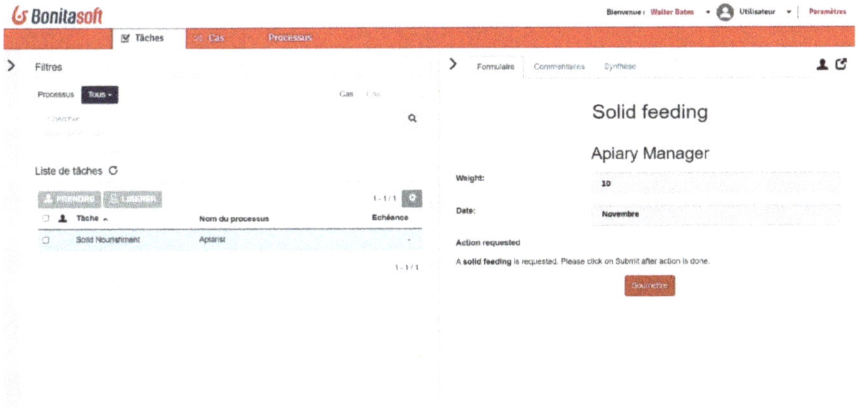

Figure 12. Simulation using Bonita software.

4. Discussion

The work proposed in this paper presents some methodological achievements in the field of beehive keeping control thanks to electronic data and remains in progress. It included the data chain from information obtained using IoT devices, workflows in the form as BPMN models, and triggering events for those models based on patterns detected in incoming data. The model-to-model approach is under development. The model transformation is not automatically done yet, but testing over pilot hives in France and Lebanon provided encouraging feedback. Moreover, the final implementation of the main key performance indicators is still under discussion.

In the context of driving and automating model transformation, authors identified several approaches, such as model driven development (MDD) within the framework for modeling simulation (MDD4MS) [34], the model driven service architecture (MDSEA [61]), the model driven interoperability for system engineering (MDISE [28]), and in this domain, "Beeomatics" [31], that can guide conceptual domain models to formal and executable models. The targeted formal models can be either used for simulation or execution of the models. As a result, the framework will integrate a transformation of BPMN models into execution and simulation models based on previous research in this domain to simulate and orchestrate/execute process models.

In any case, the interoperability will still be considered as crucial either for data or models since data can be heterogeneous, and process can integrate different data sources.

The interoperability of processes will aim to make various processes work better together. These processes will clearly define in which order services (functions) are to be executed according to the business rules [62].

As a first perspective, models are required to clearly capture the decision to be taken. For instance, the Object Management Group offers several languages that are compliant and complement BPMN onto specific features. Thus, let us mention Decisional Model Notation (DMN—OMG 2015) [63], allowing to model and represent decision and business rules within an organization. This notation is usable with BPMN and intended to be usable by businesspeople and technical people as well. In the frame of the management of a beekeeping exploitation, it will be interesting to build a set of business rules fully readable by business users and combine this kind of language with BPMN allowing to design, describe and automatize decision within a process. The DMN model will be used to pave the way and better formalize decisions in the target execution model. Moreover, the automatic model discovery and process mining will be considered as a path to capitalize on process models from historical data regarding domain practices.

Then, this work will integrate some tailored simulation code to run the dynamic animation of the BPMN model, in order to help users when facing risk in the hive. It will be proposed for a didactic purpose and decision support [64]. Considering simulation as another target model, the model-to-model transformation from BPMN as a conceptual modeling language to DEVS as a simulation model specification appears promising to run simulation and provide user decision support before choosing an action to perform. We propose to inscribe these works in open-source development frames.

The system provides recommendations for actions to be carried out to optimize the monitoring, management, and production of all apiaries. These recommendations are based on macroscopic events that originate from multiple sources of information constituting the data model, and from a behavioral model (BPMN) describing the activity of management of the apiaries. The events considered by the model can be the result of an aggregation (multicriteria) of the data collected from apiaries and from environment (weight, humidity, temperature, weather data, geographical data, video of apiaries, ...), which may be the result of business experience or learning process or both.

In addition, and as part of the future work, the apiarist will be able to choose a predefined workflow which best fits his needs or build a new one from scratch. More, beekeepers will have the choice to work in a collaborative environment where they can share their experience with local and international apiarists. The flowering and vegetation maps will be established in a collaborative way by the beekeepers of each region who will provide their own observations. The server will transform the collected data into a cartographic representation. In fact, the collaboration environment will be considered as a win-win situation since the user can benefit himself and transfer his expertise and useful information to local or even international beekeepers as well.

This cooperation will strengthen and improve practices while the system will be designed in such a way to preserve privacy for all participants. Moreover, due to multiple data sources (e.g., relative humidity, relative temperature, weather, image, sounds, etc.) and rich database system, the simulation feature will enrich the apiarist's experience and help him to manage his stock by forecasting his needs for supplies and of course will give him an eye on the production and profit. The modeling and simulation will also enhance the system's performance as far as countermeasures to be taken, based on similar previous scenarios in the past.

For now, inputs are mostly composed of hives' weights. Other information, such as weather, ambient temperature, internal hive temperature and hygrometry, and flowering, coming from the web can be added to complement the source of inputs and provide greater insight and confidence regarding the patterns of events that trigger the process models. The functioning of used IoT and extracted data format are fully known at this stage to show the usability of the approach. In the end, the final proposed application will have to allow different IoTs to inject their own data into models. Generic and interoperability mechanisms will have to be provided to allow users to exploit their own IoT and data formats without interfacing effort.

Moreover, the gamification concept will be pushed to its extreme by allowing different apiarists to share their problems, post solutions, trade ideas, and interact one with another. In fact, a rich collaborative data base collection will enhance the system's performance regarding simulation and will improve the decision making by proposing more accurate response to problems occurring inside the hives. Such a system will promote cooperation within national and international beekeepers' communities which will contribute to growth in the field.

5. Conclusions

In summary, the interest of this work can be differentiated according to the target, hobbyist beekeepers or professional beekeepers.

For the hobbyist beekeeper, process models can underpin a gamification paradigm and provide guidance for beginners. The gamification paradigm is important for this user profile.

For the professional beekeeper (with around 1000 beehives), the benefit is the optimized management of his livestock and an improvement in livestock health. In our opinion, he can expect an increase in the profitability of his bee farm in the order of 20%. The contribution of technology comes down to reducing synthetic inputs and preserving the original purity of bee products.

The advantage of the system is to provide the beekeeper with a holistic view of his farm and a precise look at each of the hives that make it up. Thus, he saves precious time for his operational organization and is supported in his decision-making. As a breeder, he knows which strains he should favor to renew his livestock. Finally, he can finely adapt the level of health treatments to the strengthen the constitution of each colony. The very precise adjustment of this level is a condition for avoiding colony losses, for minimizing the amount of artificial feeding, and for favoring the treatments accepted under organic labels.

Author Contributions: Conceptualization, C.K., A.M.C., I.K., C.Y., A.A., N.D., F.T., F.P., and G.Z.; methodology, C.K., A.M.C., I.K., C.Y., A.A., N.D., F.T., F.P., and G.Z.; software, C.K., A.M.C., I.K., C.Y., A.A., N.D., F.T., F.P., and G.Z.; validation, C.K., A.M.C., I.K., C.Y., A.A., N.D., F.T., F.P., and G.Z.; formal analysis, C.K., A.M.C., I.K., C.Y., A.A., N.D., F.T., F.P., and G.Z.; investigation, C.K., A.M.C., I.K., C.Y., A.A., N.D., F.T., F.P., and G.Z.; resources, C.Y., A.A., N.D., F.T., F.P., and G.Z.; data curation, C.K., A.M.C., and I.K.; writing—original draft preparation, C.K., A.M.C., I.K., C.Y., A.A., N.D., F.T., F.P., and G.Z.; writing—review and editing, C.K., A.M.C., I.K., C.Y., A.A., N.D., F.T., F.P., and G.Z.; visualization, C.K., A.M.C., I.K., C.Y., A.A., N.D., F.T., F.P., and G.Z.; supervision, C.K., A.M.C., I.K., C.Y., A.A., N.D., F.T., F.P., and G.Z.; project administration, C.Y., A.A., N.D., F.T., F.P., and G.Z.; funding acquisition, C.K., A.M.C., I.K., C.Y., A.A., N.D., F.T., F.P., and G.Z. All authors have read and agreed to the published version of the manuscript.

Funding: This research was partially funded by the BeePMN project from Program Hubert Curien CEDRE (France/Lebanon), grant number 4541TF.

Institutional Review Board Statement: Not Applicable.

Data Availability Statement: Publicly available datasets were used in this study. This data can be found here: https://www.kaggle.com/se18m502/bee-hive-metrics (accessed on 13 March 2021). Additionally, data was obtained from "*Connecthive*" and can be available upon request at https://connecthive.com/ (accessed on 18 March 2021).

Acknowledgments: We acknowledge the support given by "*l'Atelier du Miel*" and "*Connecthive*" which is not fully covered by the author contribution or funding sections. It includes technical and domain expert support from "*l'Atelier du Miel*", and donations from "*Connecthive*" in kind (materials used for experiments).

Conflicts of Interest: The authors declare no conflict of interest.

References

1. European Honey Bee—Apis Mellifera. Available online: http://entnemdept.ufl.edu/creatures/MISC/BEES/euro_honey_bee.htm (accessed on 24 April 2021).
2. About Honey Bees | Types, Races, and Anatomy of Honey Bees. Available online: https://www.uaex.uada.edu/farm-ranch/special-programs/beekeeping/about-honey-bees.aspx (accessed on 24 April 2021).
3. Contributions & Position Papers—European Professional Beekeepers Association. Available online: http://www.professional-beekeepers.eu/?page_id=235 (accessed on 11 June 2021).
4. Bradbear, N. Bees and their role in forest livelihoods A guide to the services provided by bees and the sustainable harvesting, processing and marketing of their products. *P Non-Wood Forest Prod.* **2009**, *19*, 194.
5. Oberreiter, H.; Brodschneider, R. Austrian COLOSS Survey of Honey Bee Colony Winter Losses 2018/19 and Analysis of Hive Management Practices. *Diversity* **2020**, *12*, 99. [CrossRef]
6. López-Uribe, M.M.; Simone-Finstrom, M. Special Issue: Honey Bee Research in the US: Current State and Solutions to Beekeeping Problems. *Insects* **2019**, *10*, 22. [CrossRef]
7. Laurino, D.; Lioy, S.; Carisio, L.; Manino, A.; Porporato, M. Vespa velutina: An Alien Driver of Honey Bee Colony Losses. *Diversity* **2020**, *12*, 5. [CrossRef]
8. Gregorc, A. Monitoring of Honey Bee Colony Losses: A Special Issue. *Diversity* **2020**, *12*, 403. [CrossRef]
9. Căuia, E.; Siceanu, A.; Vişan, G.O.; Căuia, D.; Colţa, T.; Spulber, R.A. Monitoring the Field-Realistic Exposure of Honeybee Colonies to Neonicotinoids by An Integrative Approach: A Case Study in Romania. *Diversity* **2020**, *12*, 24. [CrossRef]
10. Havard, T.; Laurent, M.; Chauzat, M.-P. Impact of Stressors on Honey Bees (Apis mellifera; Hymenoptera: Apidae): Some Guidance for Research Emerge from a Meta-Analysis. *Diversity* **2020**, *12*, 7. [CrossRef]
11. Gorecki, S.; Possik, J.; Zacharewicz, G.; Ducq, Y.; Perry, N. A Multicomponent Distributed Framework for Smart Production System Modeling and Simulation. *Sustainability* **2020**, *12*, 6969. [CrossRef]
12. Polini, W.; Corrado, A. A Unique Model to Estimate Geometric Deviations in Drilling and Milling Due to Two Uncertainty Sources. *Appl. Sci.* **2021**, *11*, 1996. [CrossRef]
13. Possik, J.; Amrani, A.; Zacharewicz, G. WIP: Co-simulation system serving the configuration of lean tools for a manufacturing assembly line. In Proceedings of the Works in Progress Symposium, WIP 2018, Part of the 2018 Spring Simulation Multiconference, SpringSim 2018, Baltimore, MD, USA, 9 July 2018. Available online: https://hal.archives-ouvertes.fr/hal-01924304 (accessed on 26 April 2021).
14. Possik, J.J.; Amrani, A.A.; Zacharewicz, G. Development of a co-simulation system as a decision-aid in Lean tools implementation. In Proceedings of the 50th Computer Simulation Conference, Society for Computer Simulation International, San Diego, CA, USA, 9–12 July 2018; pp. 1–12.
15. Gates, B.N. *The Temperature of the Bee Colony*; U.S. Department of Agriculture: Washington, DC, USA, 1914. [CrossRef]
16. Munaye, Y.Y.; Juang, R.-T.; Lin, H.-P.; Tarekegn, G.B.; Lin, D.-B. Deep Reinforcement Learning Based Resource Management in UAV-Assisted IoT Networks. *Appl. Sci.* **2021**, *11*, 2163. [CrossRef]
17. Cecchi, S.; Terenzi, A.; Orcioni, S.; Spinsante, S.; Primiani, V.; Moglie, F.; Ruschioni, S.; Mattei, C.; Riolo, P.; Isidoro, N. Multi-sensor platform for real time measurements of honey bee hive parameters. *IOP Conf. Ser. Earth Environ. Sci.* **2019**, *275*, 012016. [CrossRef]
18. Magnier, B.; Gabbay, E.; Bougamale, F.; Moradi, B.; Pfister, F.; Slangen, P. Multiple honey bees tracking and trajectory modeling. In Proceedings of the Multimodal Sensing: Technologies and Applications, Munich, Germany, 6–27 June 2019; International Society for Optics and Photonics: Bellingham, DC, USA, 2019; Volume 11059, p. 110590Z. [CrossRef]
19. Buchanan, G.; Tashakkori, R. A Web-App for Analysis of Honey Bee Hive Data. In Proceedings of the 2019 SoutheastConference, Huntsville, AL, USA, 11–14 April 2019; pp. 1–6.
20. Abu, E.S. The use of smart apiculture management system. *Asian J. Adv. Res.* **2020**, *5*, 6–16.
21. Debauche, O.; Moulat, M.E.; Mahmoudi, S.; Boukraa, S.; Manneback, P.; Lebeau, F. Web Monitoring of Bee Health for Researchers and Beekeepers Based on the Internet of Things. *Procedia Comput. Sci.* **2018**, *130*, 991–998. [CrossRef]
22. Buchmann, S.; Thoenes, S. The Electronic Scale Honey Bee Colony as a Management and Research Tool. *Bee Sci.* **1990**, *1*, 40–47.
23. Roy, C. Suivi de la Masse des Ruches: Intérêts Pour L'apiculteur et Pour le Vétérinaire Clinicien. 2019. Available online: https://www.researchgate.net/publication/337821147_Suivi_de_la_masse_des_ruches_interets_pour_l%27apiculteur_et_pour_le_veterinaire_clinicien (accessed on 8 March 2021).
24. Meikle, W.G.; Rector, B.G.; Mercadier, G.; Holst, N. Within-day variation in continuous hive weight data as a measure of honey bee colony activity. *Apidologie* **2008**, *39*, 694–707. [CrossRef]
25. Human, H.; Brodschneider, R.; Dietemann, V.; Dively, G.; Ellis, J.D.; Forsgren, E.; Fries, I.; Hatjina, F.; Hu, F.-L.; Jaffé, R.; et al. Miscellaneous standard methods for Apis mellifera research. *J. Apic. Res.* **2013**, *52*, 1–53. [CrossRef]
26. Seeley, T.D.; Visscher, P.K. Survival of honeybees in cold climates: The critical timing of colony growth and reproduction. *Ecol. Entomol.* **1985**, *10*, 81–88. [CrossRef]
27. Evans, S.K. Electronic beehive monitoring—Applications to research. In Proceedings of the 12th International Symposium of the ICP-PR Bee Protection Group, Ghent, Belgium, 15–17 September 2015; pp. 121–129.
28. Meikle, W.G.; Weiss, M.; Stilwell, A.R. Monitoring colony phenology using within-day variability in continuous weight and temperature of honey bee hives. *Apidologie* **2016**, *47*, 1–14. [CrossRef]

29. Anand, N.; Raj, V.B.; Ullas, M.S.; Srivastava, A. Swarm Detection and Beehive Monitoring System using Auditory and Microclimatic Analysis. In Proceedings of the 2018 3rd International Conference on Circuits, Control, Communication and Computing (I4C), Bangalore, India, 3–5 October 2018; pp. 1–4. [CrossRef]

30. Beehive Metrics. Available online: https://kaggle.com/se18m502/bee-hive-metrics (accessed on 23 April 2021).

31. Beeomatics, Un Système de Suivi de Rucher. Available online: https://connecthive.com/ (accessed on 25 March 2021).

32. Deterding, S.; Dixon, D.; Khaled, R.; Nacke, L. From game design elements to gamefulness: Defining "gamification". In Proceedings of the 15th International Academic MindTrek Conference: Envisioning Future Media Environments, Tampere, Finland, 30 September 2011; Association for Computing Machinery: New York, NY, USA, 2011; pp. 9–15. [CrossRef]

33. Sailer, M.; Hense, J.U.; Mayr, S.K.; Mandl, H. How gamification motivates: An experimental study of the effects of specific game design elements on psychological need satisfaction. *Comput. Hum. Behav.* **2017**, *69*, 371–380. [CrossRef]

34. Cetinkaya, D.; Verbraeck, A.; Seck, M.D. Model transformation from BPMN to DEVS in the MDD4MS framework. In Proceedings of the SpringSim, Orlando FL, USA, 30 March 2012.

35. Zacepins, A.; Brusbardis, V.; Meitalovs, J.; Stalidzans, E. Challenges in the development of Precision Beekeeping. *Biosyst. Eng.* **2015**, *130*, 60–71. [CrossRef]

36. Gisder, S.; Genersch, E. Special Issue: Honey Bee Viruses. *Viruses* **2015**, *7*, 5603–5608. [CrossRef] [PubMed]

37. Smart, M.; Otto, C.; Cornman, R.; Iwanowicz, D. Using Colony Monitoring Devices to Evaluate the Impacts of Land Use and Nutritional Value of Forage on Honey Bee Health. *Agriculture* **2018**, *8*, 2. [CrossRef]

38. Martinello, M.; Manzinello, C.; Borin, A.; Avram, L.E.; Dainese, N.; Giuliato, I.; Gallina, A.; Mutinelli, F. A Survey from 2015 to 2019 to Investigate the Occurrence of Pesticide Residues in Dead Honeybees and Other Matrices Related to Honeybee Mortality Incidents in Italy. *Diversity* **2020**, *12*, 15. [CrossRef]

39. Arnia's Hive Monitoring. Available online: http://www.arnia.co.uk/product-beekeepers/ (accessed on 14 June 2021).

40. ApisProtect. 2020. Available online: https://apisprotect.com/ (accessed on 14 June 2021).

41. BeeCare—Beehive SMS Scale. Available online: http://www.beecare.io/beehivesmsscale/?utm_source=Klix.ba&utm_medium=Clanak (accessed on 14 June 2021).

42. Apiary Excellence | BuzzTech. Available online: https://buzztech.nz/index.html (accessed on 14 June 2021).

43. Nuu-Bee, T. Solutionbee. Available online: https://solutionbee.com/ (accessed on 14 June 2021).

44. Broodminder. Available online: https://broodminder.com/ (accessed on 15 June 2021).

45. OSBeehives | BuzzBox Hive Health Monitor & Beekeeping App. Available online: https://www.osbeehives.com (accessed on 15 June 2021).

46. Moawad, A.; Hartmann, T.; Fouquet, F.; Nain, G.; Klein, J.; Traon, Y.L. Beyond discrete modeling: A continuous and efficient model for IoT. In Proceedings of the 2015 ACM/IEEE 18th International Conference on Model Driven Engineering Languages and Systems (MODELS), Ottawa, ON, Canada, 30 September–2 October 2015; pp. 90–99. [CrossRef]

47. Allison, P. Can A Beehive Get Too Hot or Too Cold? *Allisons Apiaries* 2018. Available online: https://allisonsapiaries.com/ideal-beehive-temperature-bees-honey/ (accessed on 25 March 2021).

48. L'Atelier Du Miel—Pure, Raw Honey, 100% Natural. Shop Now! Available online: https://www.atelierdumiel.com/ (accessed on 25 March 2021).

49. OMG Business Process Model and Notation (BPMN) Version 2.0.2. 2013. Available online: https://www.omg.org/spec/BPMN/2.0.2/PDF (accessed on 27 March 2021).

50. BPMNPoster—www.bpmb.de. Available online: http://www.bpmb.de/index.php/BPMNPoster (accessed on 23 April 2021).

51. Possik, J.; D'Ambrogio, A.; Zacharewicz, G.; Amrani, A.; Vallespir, B. A BPMN/HLA-Based Methodology for Collaborative Distributed DES. In Proceedings of the 2019 IEEE 28th International Conference on Enabling Technologies: Infrastructure for Collaborative Enterprises (WETICE), Napoli, Italy, 12–14 June 2019; pp. 118–123. [CrossRef]

52. Process-Aware Information Systems: Bridging People and Software Through Process Technology | Wiley. Available online: https://www.wiley.com/en-us/Process+Aware+Information+Systems%3A+Bridging+People+and+Software+Through+Process+Technology-p-9780471663065 (accessed on 23 April 2021).

53. Press, T.M. Workflow Management | The MIT Press. Available online: https://mitpress.mit.edu/books/workflow-management (accessed on 23 April 2021).

54. Workflow Management Coalition Glossary and Terminology. Available online: http://www.aiai.ed.ac.uk/project/wfmc/ARCHIVE/DOCS/glossary/glossary.html (accessed on 23 April 2021).

55. Frank, L.; Roller, D. *Production Workflow: Concepts and Techniques*; Prentice Hall: Upper Saddle River, NJ, USA, 2000; ISBN 0-13-021753-0.

56. Zacharewicz, G.; Frydman, C.; Giambiasi, N. G-DEVS/HLA Environment for Distributed Simulations of Workflows. *Simulation* **2008**, *84*, 197–213. [CrossRef]

57. Ougaabal, K.; Zacharewicz, G.; Ducq, Y.; Tazi, S. Visual Workflow Process Modeling and Simulation Approach Based on Non-Functional Properties of Resources. *Appl. Sci.* **2020**, *10*, 4664. [CrossRef]

58. Long Short-Term Memory | Neural Computation | MIT Press. Available online: https://direct.mit.edu/neco/article/9/8/1735/6109/Long-Short-Term-Memory (accessed on 23 April 2021).

59. Kingma, D.P.; Ba, J. Adam: A Method for Stochastic Optimization. *arXiv* **2017**, arXiv:14126980. Available online: http://arxiv.org/abs/1412.6980 (accessed on 23 April 2021).

60. Getting Things Done®—David Allen's GTD®Methodology. Available online: https://gettingthingsdone.com/ (accessed on 23 April 2021).
61. Bazoun, H.; Bouanan, Y.; Zacharewicz, G.; Ducq, Y.; Boye, H. Business process simulation: Transformation of BPMN 2.0 to DEVS models (WIP). In Proceedings of the Symposium on Theory of Modeling & Simulation—DEVS Integrative, San Diego, CA, USA, 13–16 April 2014; Volume 46, pp. 1–7.
62. Chen, D.; Daclin, N. Framework for enterprise interoperability. In *Proceedings of the IFAC Workshop EI2N*; ISTE: London, UK, 2006; pp. 77–88.
63. OMG Decision Model and Notation Version 1.0. Object Managemnet Group (OMG). 2015. Available online: https://www.omg.org/spec/DMN/1.0/PDF (accessed on 27 March 2021).
64. Pfister, F.; Chapurlat, V.; Huchard, M.; Nebut, C.; Wippler, J.-L. A proposed meta-model for formalizing systems engineering knowledge, based on functional architectural patterns. *Syst. Eng.* **2012**, *15*, 321–332. [CrossRef]

 *diversity*

Article

Austrian COLOSS Survey of Honey Bee Colony Winter Losses 2018/19 and Analysis of Hive Management Practices

Hannes Oberreiter and Robert Brodschneider *

Institute of Biology, University of Graz, Universitätsplatz 2, 8010 Graz, Austria; hoberreiter@gmail.com
* Correspondence: robert.brodschneider@uni-graz.at; Tel.: +43-316-380-5602

Received: 7 February 2020; Accepted: 10 March 2020; Published: 13 March 2020

Abstract: We conducted a citizen science survey on overwinter honey bee colony losses in Austria. A total of 1534 beekeepers with 33,651 colonies reported valid loss rates. The total winter loss rate for Austria was 15.2% (95% confidence interval: 14.4–16.1%). Young queens showed a positive effect on colony survival and queen-related losses. Observed queen problems during the season increased the probability of losing colonies to unsolvable queen problems. A notable number of bees with crippled wings during the foraging season resulted in high losses and could serve as an alarm signal for beekeepers. Migratory beekeepers and large operations had lower loss rates than smaller ones. Additionally, we investigated the impact of several hive management practices. Most of them had no significant effect on winter mortality, but purchasing wax from outside the own operation was associated with higher loss rates. Colonies that reported foraging on maize and late catch crop fields or collecting melezitose exhibited higher loss rates. The most common *Varroa destructor* control methods were a combination of long-term formic acid treatment in summer and oxalic acid trickling in winter. Biotechnical methods in summer had a favourable effect on colony survival.

Keywords: *Apis mellifera*; varroa control; colony losses; forage; beekeeping; citizen science; overwintering; monitoring

---

1. Introduction

Apis mellifera, the western honey bee, is not only an important pollinator for wild flowers but also crucial for the pollination of the world agricultural production. It was calculated in 2005 that the estimated economic value of pollinating insects in the European Union equals 14.2 billion euro [1]. The threats for honey bees are increasingly well studied and understood and range from abiotic stressors, such as pesticides, to biotic stressors [2–4]. The latter include parasites, pathogens, and pests, like the parasitic mite *Varroa destructor* or monoculture plantings that influence the quality and richness of forage sources [5]. Further, synergistic effects of single factors may add up to a threat that is greater than the sum of its individual factors. Beekeepers try their best to support their colonies by optimizing their hive management to help them cope with the environmental conditions they are facing. Comparisons of hive management practices of large data sets collected from beekeepers have demonstrated different strategies and consequently different overwintering success [6–9]. For example, efficient treatments of colonies against varroa mite are necessary, as this parasite is known to reduce winter survival [10–12].

Honey bee monitoring via citizen scientist beekeepers or crowdsourcing has been carried out for a couple of years in Austria [13] and other countries [6,7]. It helps to identify potential risk factors and provides a data base to gain a better understanding of honey bee colony losses. Thanks to the nonprofit honey bee research association COLOSS (Prevention of honey bee COlony LOSSes) [14], monitoring of honey bee colonies is done in various European and some non-European countries following international standards [15]. This facilitates the comparison between countries and a joint identification of risk factors [16,17]. The analysis of this data has already revealed successful practices for colony management and identified unfavourable biotic factors, such as crops which negatively influence colony survival [18]. In Austria, a substantial data set, which has been acquired over several years, has already been used to investigate the effect of weather [19] and land use [20] on colony losses.

In contrast to previous COLOSS publications [7,16] that have combined loss rates from countries with quite different environmental conditions and hence are limited in analyses of regional risk factors, we present a complete examination of the data collected within COLOSS for Austria. This complements the last comprehensive risk analysis on Austrian honey bee colony winter mortality published a while ago [13]. We present a study using a large data set obtained by crowdsourcing for winter 2018/2019 in Austria and a substantially improved methodology. The aim is to report internationally comparable loss rates, to identify factors which have negative or positive effects on honey bees, and to reappraise different hive management techniques.

2. Materials and Methods

2.1. Survey Design and Response Rate

Our survey was based on the questions of the international COLOSS questionnaire, which has been translated to German. The questionnaire asks for the number of honey bee colonies wintered and the number of colonies lost with three possible categories (colony dead or reduced to a few hundred bees, lost due to natural disaster (i.e., flooding, falling trees, and vandalism), or alive but with unsolvable queen problems (drone laying queen or no queen at all)) [15]. The number of colonies lost due to natural disaster was not included in the total loss calculation and risk analysis because it is not directly related to biological (i.e., age of queen bees) or operational risk factors in the survey.

Questions on hive management practices are also part of the COLOSS international survey and were used unmodified. These questions concerned topics like the number of the wintered colonies that had a new queen (born and mated 2018), observed queen problems in colonies during foraging season compared to previous seasons, certified organic beekeeping, queens bred from varroa tolerant/resistant stock, hives fabricated from synthetic materials, insulated hives, screened bottom boards, purchase of wax from outside of the own operation (a measure of professionalism and wax quality), natural comb without foundation, small brood cell size, migratory beekeeping, replacement of brood frames (in relative percentage categories), notice of bees with crippled/deformed wings (often, seldom, none, and do not know), and foraging crops perceived by the beekeeper. As for the foraging crops, the participants were asked if the majority of their bee colonies had significantly foraged on one or more of the asked crops (only crops relevant for Austria were listed). The selection of some of these operational factors followed suggestions from beekeepers in previous surveys or discussion groups and were accepted for international use in COLOSS.

An important part of the survey was dedicated to varroa control. Beekeepers could identify if and in which month they monitored varroa infestation level. Additionally, the methods and application time of varroa control were surveyed.

The estimated percentage of beekeepers participating in the survey was calculated with the number of beekeepers and colonies registered at the national beekeeping association (Table 1). Data collection was carried out via an online survey (LimeSurvey Version 3.16.1+) and a questionnaire published in a beekeeping journal and physically handed out at beekeeper meetings. This ensured that also beekeepers without an internet connection had access to the survey [15]. Overall, the survey followed the guidelines of the COLOSS Project rather closely [15]. To guarantee the protection of the participants private data, a data privacy protocol was established between the people involved in the project. All private data was removed when working on the analysis of received survey data and only used to enquire questionable data.

Table 1. Number and percentage of survey participants in relation to the amount of registered beekeepers with the national beekeeping association "Biene Österreich" (2018): Commercial beekeepers are subsumed as "not specified" due to the absence of state data.

	Beekeepers	Survey particip.	Survey (%)	Colonies	Survey Colonies	Survey (%)
Burgenland	642	35	5.5	11,530	759	6.6
Carinthia	3013	147	4.9	33,993	3784	11.1
Lower Austria	4605	387	8.4	41,414	9486	22.9
Upper Austria	8075	276	3.4	80,000	5951	7.4
Salzburg	2574	74	2.9	19,035	1586	8.3
Styria	4038	218	5.4	54,960	5706	10.4
Tyrol	2825	141	5.0	36,094	2963	8.2
Vorarlberg	1530	178	11.6	10,106	2333	23.1
Vienna	707	78	11.0	6124	1083	17.7
not specified	423	-	-	80,156	-	-
Total	**28,009**	**1534**	**5.5**	**373,412**	**33,651**	**9.0**

2.2. Data Validation and Error Control

If beekeepers did participate via paper questionnaire, the data was manually transferred to a Microsoft Excel file, where all survey data was collected. Automatic checks with simple formulas in Excel were used to minimize processing errors and to highlight possible invalid responses, i.e., more colonies lost than existent, as described in References [15,16]. These contradictory entries or multiple entries of the same beekeeper were removed. The survey in the beekeeping journal did not cover all the questions that were asked in the internet survey, and omitted answers also led to a reduced number of responses for several questions. If the response for one factor was not enough for statistical analysis, the data was not used. The participants did not disclose the exact location of their apiaries, only the rough area of their main wintering apiary, i.e., district and zip code at minimum. Coordinates for plotting were generated with a batch geolocation finder and, if not feasible, by a manual search. The resulting locations were plotted on a district map and tested for correct assignment to districts as given by the participants in the survey, to minimize incorrect geolocations. The estimation of elevation for the apiary locations was carried out with the topography model SRTM-3v4 via a web service [21].

2.3. Statistical Analysis

The statistical software R [22] and the package ggplot2 [23] were used for data analysis and the construction of graphics and plots. The code is available on GitHub (https://github.com/HannesOberreiter/coloss_honey_bee_colony_losses_austria and version 1.0 is archived [24]) under an open source MIT license. Calculation and estimation of loss rates and corresponding confidence intervals (CI) were computed with a quasibinominal generalized linear model (GZLM) link "logit" function [15]. These calculated estimates are plotted as boxed error bars, where the box represents the 95% CI.

The majority of analyses was done as a single factor with two possible different groups, i.e., yes and no questions. To identify significant differences, the confidence intervals between the variables (factors) were compared. If the confidence intervals did not overlap, we counted that as a significant difference. If the overlap was on a small margin and it was a comparison between two groups, the null deviance minus the residual deviance in the model was tested. If it significantly differed from zero, the analyzed factor in the model had a significant influence on colony survival (ANOVA with χ^2 deviance, $p < 0.05$).

The analysis of young queens was only performed for participants with valid answers for the number of colonies and young queens at the onset of winter. The percentage was calculated with declared colonies and divided by young queens. The total young queen population was calculated with these valid answers by summarising all colonies and by dividing them by the total number of young queens going into the winter.

For the analysis of the combination of different varroa control methods, a usage histogram for each method was generated and grouped into spring, summer, and winter. After that, a vector of all applied methods with at least 15 answers and without drone brood removal was generated to minimize calculation time and to generate statistically relevant results. To calculate all possible combinations inside the given vector, the R function combn [25] was used to generate a matrix with up to a maximum of three different methods per row. Then, each row with less than 15 participants for the particular combination was dropped before calculating the loss rate as described before.

The plotted maps were made with R [22] and shapefiles (https://github.com/ginseng666/ GeoJSON-TopoJSON-Austria by Floo Perlot, latest commit 2017-11-06) under creative common licence. Aggregation of apiaries on the map was conducted with k-means cluster search method. To improve the accuracy in areas with lower density, the resulting clusters, which included only a single apiary or a low number of apiaries and high within cluster sum of squares, were removed and the original geolocation of the affected apiaries were used for plotting.

3. Results

3.1. Survey Data

The distribution of the number of colonies managed by beekeepers showed that most of the participating beekeepers (50.8%) had between 1–10 colonies (Figure 1A). On the contrary, only 1.7% of the beekeepers had more than 150 colonies but owned 21.5% of the colonies in the survey (Figure 1B). The distribution of losses per individual operation showed a positive skew with more than half (51.4%) of the participating beekeepers loosing between 0–10% of their colonies over the winter period and only 30.3% loosing more than 20% (Figure 1C). The approximate location of the main winter apiary (Figure 1D) showed a nationwide coverage all over Austria, with some areas more dominant than others, which could be traced back to geographical inaccessible areas like mountain ranges. Most participants did use the online form (n = 1378, 90%), while the paper form (n = 93, 6%) and the questionnaire from the beekeeping journal (n = 63, 4%) only accounted for 10% of the responses. The loss rate was not significantly different between online form users 15.4% (95% CI: 14.5–16.3%, n = 1378) and users who responded via paper 13.2% (95% CI: 10.4–16.5%).

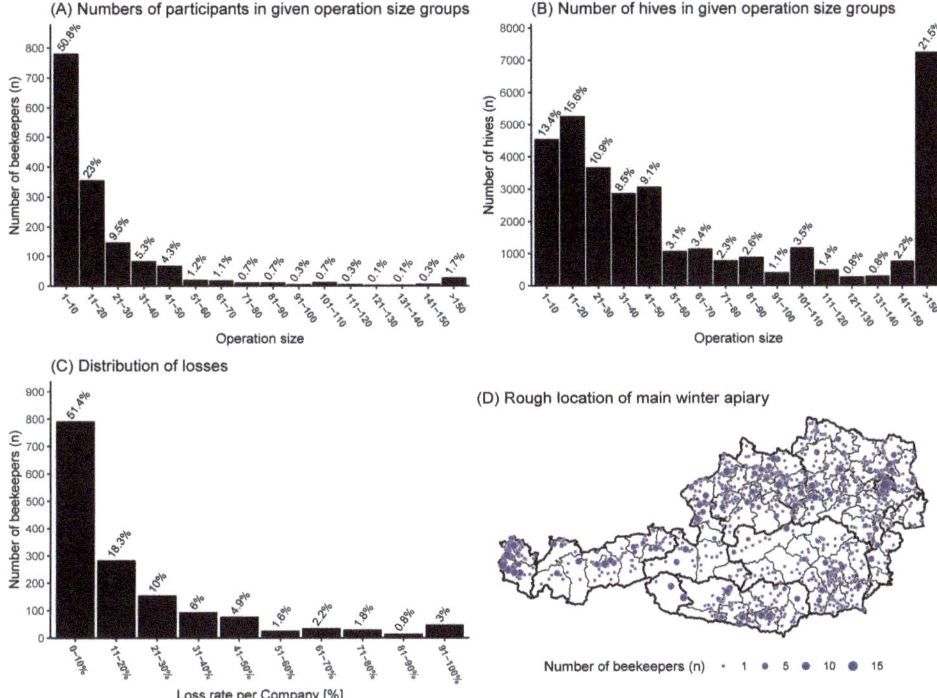

Figure 1. (**A**,**B**) Operation size distribution in the survey: The operation size is the number of colonies owned by the individual beekeeper or beekeeping operation. (**C**) Distribution of losses in the survey, grouped into 10% loss rate groups. (**D**) Approximate location of the main winter apiary location: dot size represents number of apiaries in this area.

3.2. Loss Rate Overview

The total colony mortality over the winter period of 2018/2019 in Austria equals 15.2% (95% CI: 14.4–16.1%) (Figure 2A). The survey received valid answers from 1534 beekeepers, which stated that they wintered 33,651 colonies and had lost 5293 colonies over the winter (Table 1). Among these lost colonies, 1304 colonies were reported to struggle due to unsolvable queen-related problems and were counted as losses. Additionally, 60 colonies were lost due to natural disasters, which are not included in our risk analysis. The loss rate of the state Carinthia with 11.5% (95% CI: 9.5–14.0%) is significantly lower than the Austrian average. Burgenland had the lowest loss rate with 9.9% (95% CI: 6.0–15.7%) (Figure 2B) but also had the lowest number of participants (n = 35) and a wide confidence interval. Vienna had the highest loss rate but also the widest confidence interval with 19.6% (95% CI: 14.8–25.4%). The loss rates at the district level were highest for Waidhofen an der Ybbs, Lower Austria with 65.0% (95% CI: 48.6–79.9%, n = 7) and lowest for Tamsweg, Salzburg 0.0% (95% CI: 0.0–100.0%, n = 6).

The elevation of the main winter apiary location showed significantly lower loss rates for higher elevations (601–800 m: 13.0% (95% CI: 11.3–15.0%), >800 m: 11.8% (95% CI: 9.9–14.0%)) than the middle groups (201–400 m: 17.0% (95% CI: 15.4–18.9%), 401–600 m: 16.6% (95% CI: 15.0–18.2%)). No significant difference can be seen if compared to the lowest elevation group (0–200 m) with 15.4% (95% CI: 12.8–18.4%) (Figure 3).

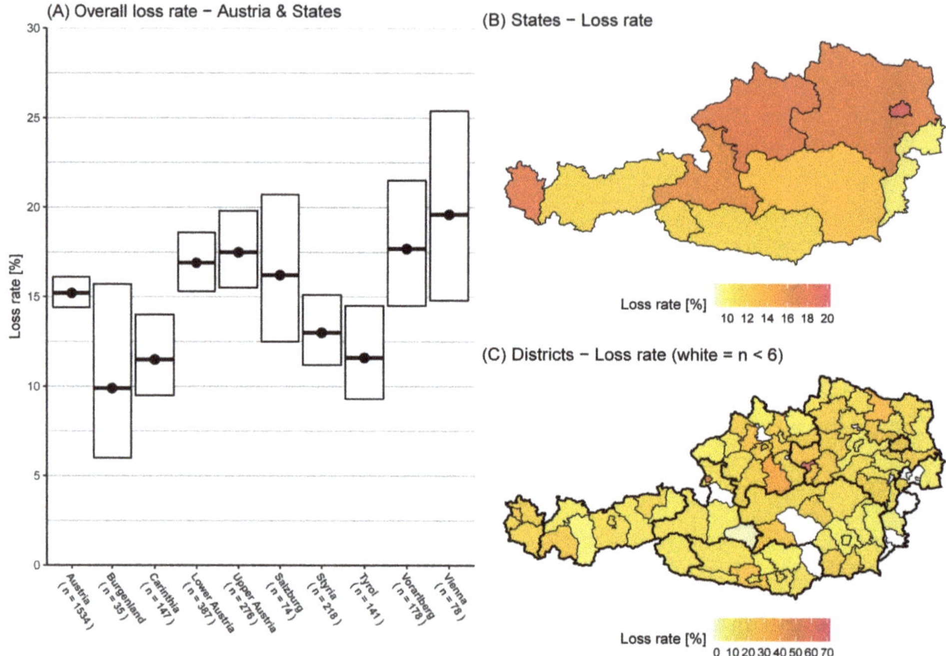

Figure 2. (**A**) Austria and its state winter honey bee colony loss rates (and 95% CI) over the winter period of 2018/2019. (**B**) Map with loss rates colour coded on the Austrian state level. (**C**) Map with loss rates colour coded on the district level: white spots are *n* < 6.

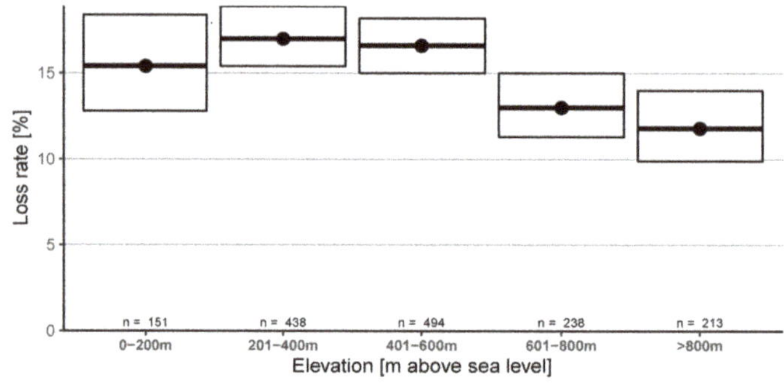

Figure 3. Winter honey bee colony loss rates (and 95% CI) grouped into five different elevation above sea level categories.

3.3. Queen Management

Colonies, which were still alive post winter but had queen problems that could not be solved by the beekeeper (drone laying queens or no queen at all) were counted as queen-related losses. This accounted for less than half of the combined losses, with an overall loss rate in Austria of 3.9% (95% CI: 3.6–4.1%). Lower Austria, 4.8% (95% CI: 4.3–5.4%), showed a significantly higher queen-related loss rate than the Austrian average (Figure 4A).

The distribution of the survey data showed that most beekeepers (85.1%) did renew more than a quarter of their queens (Figure 4B) and only 14.9% exchanged 0–25%. These result in a young queen percentage in the winter colonies population of about 52.3%. Loss rate without queen-related losses in relation to the relative number of exchanged young queens, categorised in four groups, showed a significant higher loss rate for the first two groups with a lower number of renewed queens (0–25%: 17.1% (95% CI: 14.2–20.5%), 26–50%: 12.7% (95% CI: 11.4–14.3%)). The third group (51–75%) showed a significantly lower loss rate 10.1% (95% CI: 8.8–11.7%) compared to the first group, and the last group (75–100%) with most renewed queens had the lowest loss rate with 9.2% (95% CI: 7.4–11.3%) and was significantly lower than the first two groups (Figure 4C). Colony losses due to queen-related problems, using the same four categories of young queens, showed again a significantly higher loss rate for the first group with a 0–25% exchange rate and with 5.6% (95% CI: 4.7–6.6%) loss rate than the other three groups (26–50%: 3.8% (95% CI: 3.5–4.3%), 51–75%: 3.7% (95% CI: 3.3–4.1%), and 76–100%: 3.3% (95% CI: 2.7–3.9%)) (Figure 4D).

Participants were asked to what extent they observed queen problems in their colonies during the foraging season compared to previous seasons (four categories: more often, normal, more rare, and do not know). Only 7.8% of the participants stated that problems occurred more often, and 80.1% reported less or a normal experience (Figure 5A). There was no significant difference between these categories to loss rate without queen-related losses (Figure 5B). If looking at the queen-related losses in relation to the given categories, the category, "more often" had a significantly higher loss rate with 6.0% (95% CI: 4.8–7.3%) than "more rare" with 3.7% (95% CI: 3.1–4.3%) (Figure 5C).

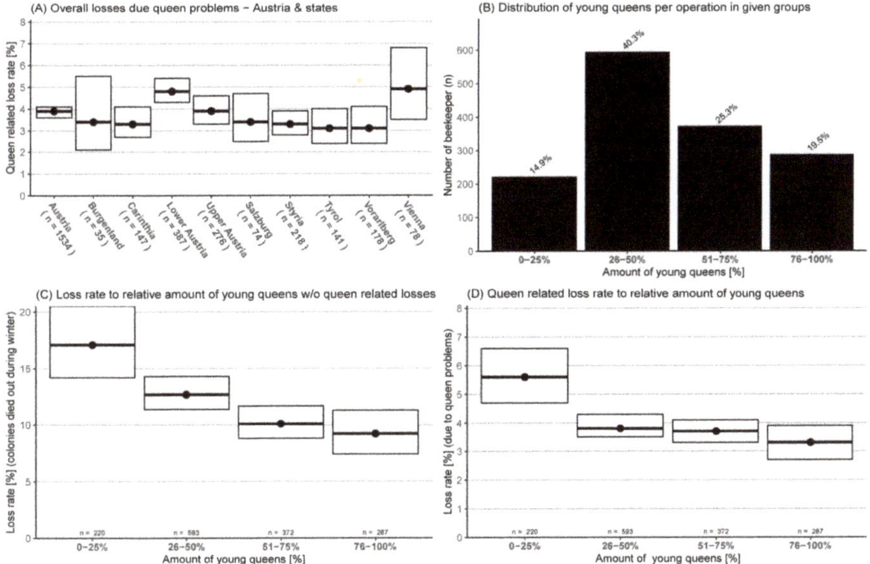

Figure 4. (**A**) Queen-related losses (and 95% CI) for Austria and its states. (**B**) Distribution of survey participants who replaced their old queens in four categories. (**C**) Categorized relative number of young queens and their winter honey bee colony loss rates (and 95% CI) without queen-related losses. (**D**) Categorized relative number of young queens and queen-related loss rate (and 95% CI).

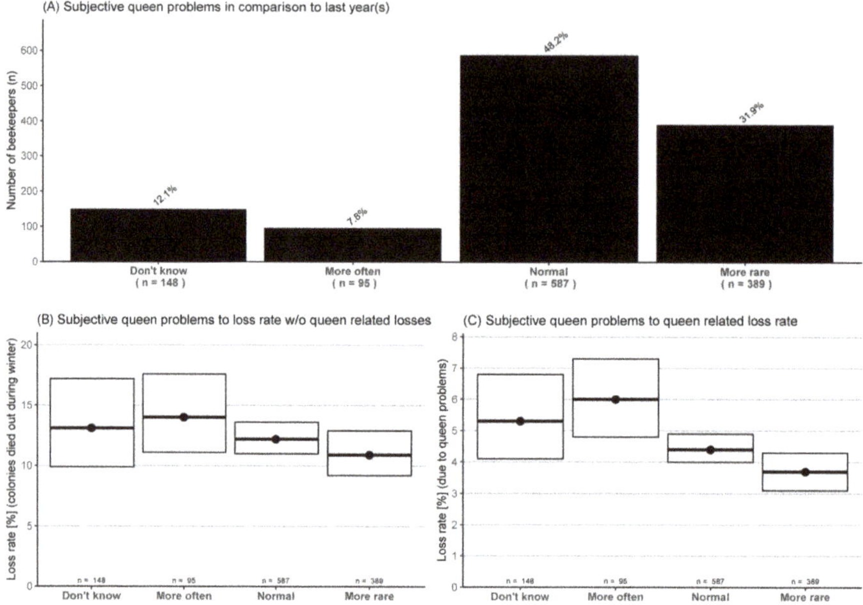

Figure 5. (**A**) Rating of observed queen problems during the foraging season in four categories compared to previous season(s). (**B**) Winter honey bee colony loss rates (and 95% CI) without queen-related losses to queen problem occurrences. (**C**) Queen-related loss rate (and 95% CI) related to queen problem occurrences.

3.4. Hive Management Practices

Nine operational hive management practices were provided with "yes", "no", or "uncertain" as possible options and compared as a single factor to colony mortality (Figure 6). Participants who did migrate their colonies had a significantly lower loss rate 11.3% (95% CI: 10.1–12.7%) than the ones who did not migrate their colonies 17.3% (95% CI: 16.3–18.5%) (Figure 6B). If beekeepers purchased wax from outside their own operation, they had a significantly higher loss rate 17.4% (95% CI: 15.8–19.2%) than participants with their own wax 14.0% (95% CI: 13.0–15.1%) (Figure 6G). No other significant effect of an operational factor on winter loss rate was found (Figure 6). The highest number of participants answering "uncertain" ($n = 124$, 9.3%) was recorded for the question whether small brood cell size was used.

Another question concerned the amount of old brood frames that were exchanged in the previous summer (Figure 7). The four categories in the survey were compared, which provided us with the relative amount of renewed brood frames by the participants. Higher exchange rates (>30%) showed a trend to lower loss rate but with no significant difference between the categories (Figure 7).

Reports categorized by operation size showed a significantly higher loss rate for beekeepers with 1–25 colonies (17.5%, 95% CI: 16.0–19.1%) over participants with 26–50 colonies (13.0%, 95% CI: 11.4–14.7%) and over 50 colonies (14.5%, 95% CI: 13.3–15.9%) (Figure 8).

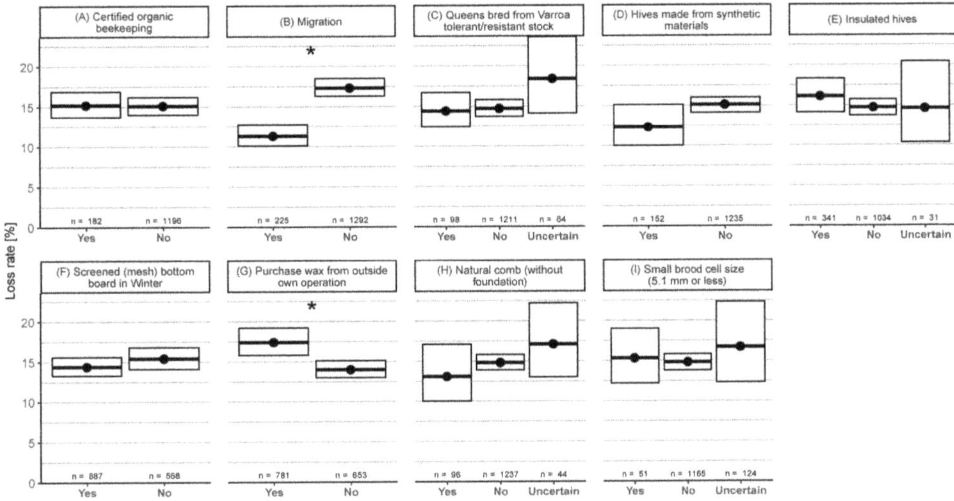

Figure 6. Winter honey bee colony loss rates (and 95% CI) for nine different hive management practices with three possible answers: "yes", "no", or "uncertain". These were compared as single factors to the loss rates. The category "uncertain" was removed from the plot when there were less than 30 answers.

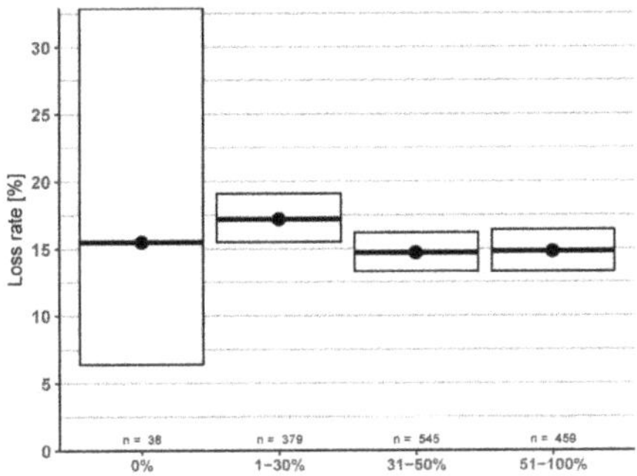

Figure 7. Comparison between relative amount of exchange rate of old brood frames in four given percentage groups and corresponding winter honey bee colony loss rates (and 95% CI).

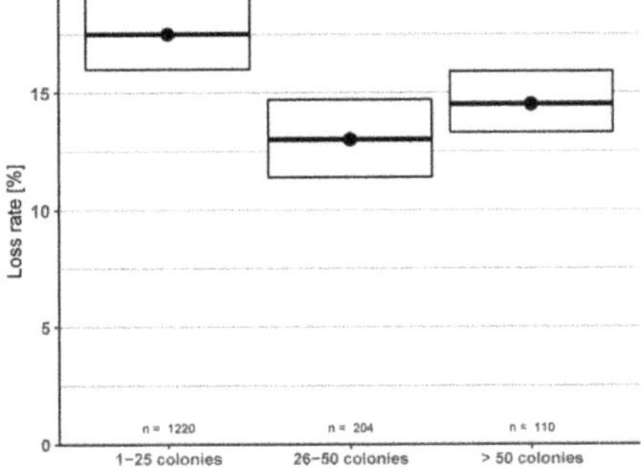

Figure 8. Categorized operation size groups and their winter honey bee colony loss rates (and 95% CI): The number of reported colonies before the onset of winter (2018) were used for the grouping.

3.5. Forage as Risk Factor

Participants were asked in the survey if their bees foraged on the following sources in 2018: oilseed rape (*Brassica napus*), maize (*Zea mays*), sunflower (*Helianthus annuus*), late catch crop, honeydew, and melezitose.

Significantly higher loss rates were observed with melezitose (yes: 17.4% (95% CI: 15.4–19.5%), no: 14.1% (95% CI: 13.1–15.3%)) (Figure A1F) and late catch crop (yes: 15.9% (95% CI: 14.5–17.3%, χ^2 = 29.2, $p < 0.05$), no: 13.4% (95% CI: 12.0–15.0%)) (Figure A1D), and colonies with bees reported foraging in maize fields (yes: 18.1% (95% CI: 15.0–21.8%), no: 13.8% (95% CI: 12.7–14.9%)) (Figure A1B).

No difference was observed for rapeseed (yes: 16.5% (95% CI: 14.7–18.6%), no: 14.6% (95% CI: 13.5–15.8%)) (Figure A1A), sunflower (yes: 15.9% (95% CI: 13.9–18.1%), no: 14.9% (95% CI: 13.8–16.1%)) (Figure A1C), and honeydew (yes: 13.9% (95% CI: 12.5–15.5%), no: 15.5% (95% CI: 14.2–16.8%)) (Figure A1E).

The spatial distribution of the crops as reported by beekeepers revealed an aggregation of oil seed rape in Upper Austria and Lower Austria (Figure A1G). Most sunflower fields were present in Lower Austria (Figure A1I). Late catch crop and honeydew occurred all over Austria (Figure A1J,K). Melezitose occurred predominantly in Styria and Carinthia (Figure A1L).

3.6. Varroa Control

3.6.1. Overview

Participants reported the number of sighted bees with crippled/deformed wings in four categories. There was a significantly higher loss rate in the category "often" 36.3% (95% CI: 28.0–45.5%) than in "seldom" 16.8% (95% CI: 15.4–18.3%). The category "seldom" was also significantly higher than "none" with 13.5% (95% CI: 12.5–14.7%) (Figure 9B).

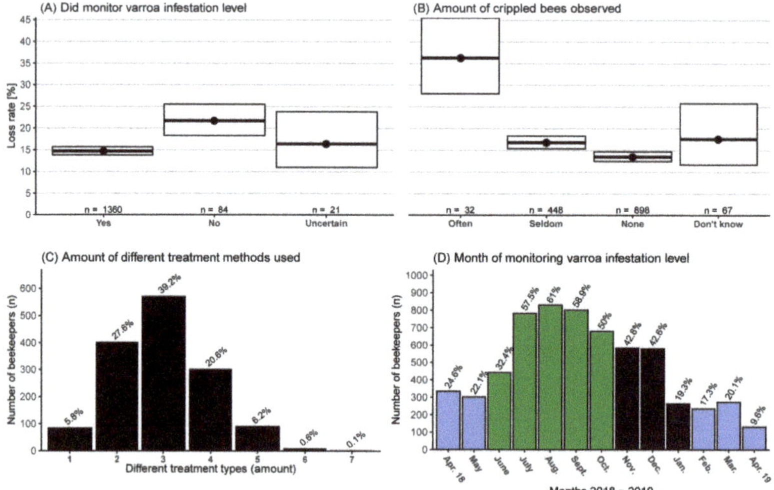

Figure 9. (**A**) Categories for if monitoring for varroa mites during the period April 2018–April 2019 was practiced and the corresponding winter honey bee colony loss rates (and 95% CI). (**B**) If and how often the participants noticed bees with crippled/deformed wings in their colonies (during summer season) in four categories and their loss rate (and 95% CI). (**C**) Distribution of how many different treatment methods (including drone brood removal) per operation were used, i.e., formic acid—short term and thymol would be two. (**D**) Histogram of which months monitoring of varroa infestation level (e.g., counting mite fall) was done and by how many beekeepers (*n* = 1360) in the survey. Spring, summer, and winter are color coded; see Figure 10.

Beekeepers who monitored varroa infestation level had a significantly lower loss rate at 14.7% (95% CI: 13.8–15.7%) than those that did not at 21.7% (95% CI: 18.3–25.5%) (Figure 9A). The monitoring of varroa infestation level was primarily performed in the months between July and October (Figure 9D).

Varroa control information was provided by 1455 participants. Most beekeepers (87.4%) used between two and four different treatment methods (Figure 9C). The most common treatment in spring (April–May) was drone brood removal (Figure 10A). In summer (June–October) the prevalent treatment was formic acid, with more participants using a long-term than short-term method (Figure 10C,D). In the autumn/winter period (November–January), oxalic acid treatment in its various forms was the most dominant (Figure 10B,E,F). Lactic acid, synthetic methods (including Amitraz, Coumaphos, and other synthetic methods), and "another method" (without synthetic methods) were the least common used treatments in the survey (Figure 10J–L).

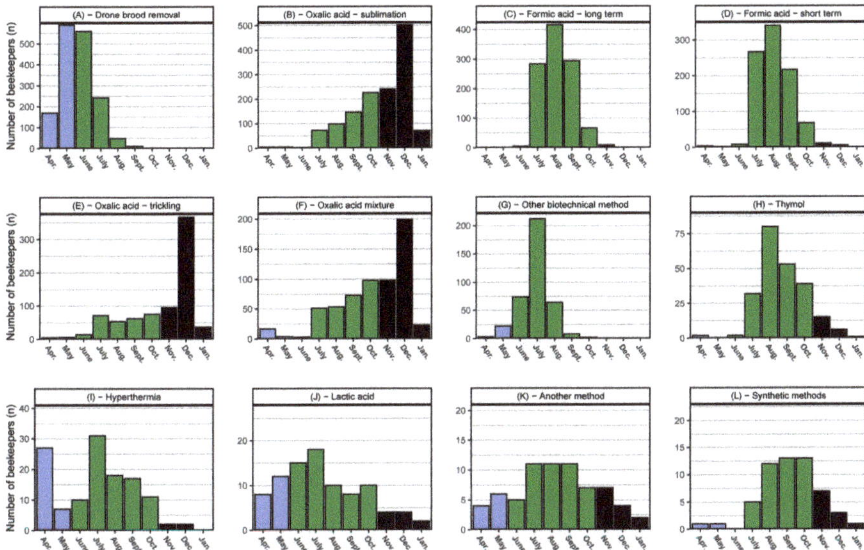

Figure 10. Varroa control methods and their month of usage as histograms starting with April 2018 and sorted by total frequencies. (**L**) "Synthetic methods" are combined and include Amitraz, Coumaphos, and "other synthetic methods"; (**G**) "other biotechnical methods" do not include drone brood removal or hypothermia; and (**K**) "another method" excludes synthetic methods. We defined April–May as spring (blue), June–October as summer (green), and November–January as autumn/winter (black). The months February, March, and April 2019 are excluded. The Y-axis has separate scales for each treatment. *n* = 1455 participants.

3.6.2. Treatment as Single Factor

The colony loss rates of operations with different treatment methods were compared. Methods were grouped into spring, summer, and winter according to the month of usage (Figure 10). Within the groups, we did not differentiate if the same treatment methods were used multiple times by any given beekeeper, only if they were performed.

In spring (April–May), four control methods were applied by at least 20 participants (Figure A2). Results of drone brood removal as a single factor is discussed in the next section. The application of the other three spring control methods (hyperthermia, other biotechnical method, and oxalic acid trickling) had no significant effect on winter loss rates.

For summer (June–October), eleven treatment methods (yes = $n > 19$) were compared (Figure A3). Significantly lower loss rates were found for participants using other biotechnical methods (13.1%, 95% CI: 11.7–14.8%) than for those participants not using them (16.2%, 95% CI: 15.2–17.4%) (Figure A3C). Participants who performed oxalic acid trickling (including oxalic acid mixtures) in summer had a significantly higher loss rate (17.4%, 95% CI: 15.7–19.4%) compared to those who did not use this treatment method (14.6%, 95% CI: 13.7–15.7%) (Figure A3H).

Only oxalic acid methods (sublimation and trickling) were applied in winter (November–January) with a minimum of 20 reports. No significant differences between the application or non-application of these methods were found (Figure A4).

3.6.3. Drone Brood Removal Combination

Drone brood removal performed only in spring, only in summer, and in both seasons did not have an effect on winter losses compared to no drone brood removal at all. Removing drone brood only in summer was significantly worse (17.6%, 95% CI: 15.3–20.3%) than drone brood removal in spring and summer (13.3%, 95% CI: 11.8–15.0%) (Figure 11).

Drone brood removal as a single factor in spring (which includes also the abovementioned beekeepers that apply this treatment in both spring and summer) resulted in a just significant effect compared to those not removing drone brood in spring: confidence intervals are overlapping (yes: 14.1% (95% CI: 12.8–15.4%), no: 16.3% (95% CI: 15.1–17.5%)) (Figure A2A), but $\chi^2 = 29.7$ ($p < 0.05$). In summer, no significant difference was found between whether beekeepers applied this control method.

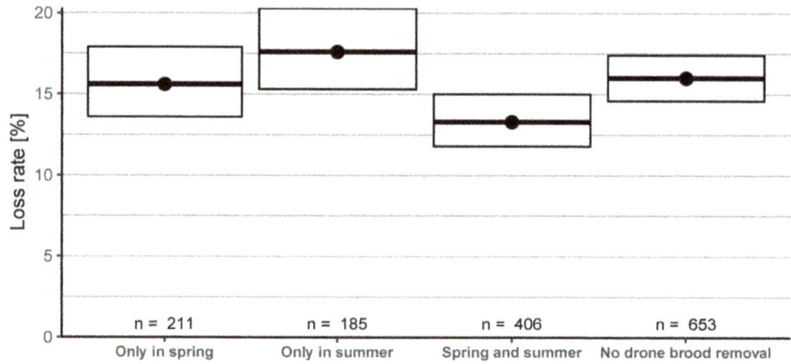

Figure 11. Winter honey bee colony loss rates (and 95% CI) in dependence of drone brood removal performed only in spring, only in summer, in spring and summer or not performed at all.

3.6.4. Treatment Combinations

To gain a more thorough understanding of the used methods, their combinations, and their success, all possible combinations of different treatment methods were calculated. How the combinations were generated and selected is described in the Materials and Methods section. The used letters and abbreviations for the generated treatment combinations as seen in Figure 12 are explained in Table 2.

The most used method combination in the survey was (A) formic acid—long term in summer—and oxalic acid—trickling in winter. It showed the smallest confidence interval and significantly lower loss rate of 10.8% (95% CI: 9.2–12.6%) than the second most frequent combination (B) formic acid—short term in summer—and oxalic acid—trickling in winter—with 16.1% (95% CI: 13.0–19.6%), combination (D) formic acid—short term in summer—and oxalic acid—sublimation in winter—with 16.6% (95% CI: 12.6–21.5%), and the combination (H) formic acid—long term in summer—and oxalic acid—sublimation in winter and summer with—20.1% (95% CI: 15.1–26.1%). The highest loss rates were observed for (L) formic acid—short term in summer—and oxalic acid—trickling in summer and winter—with a loss rate of 26.9% (95% CI: 18.6–37.3%), combination (N) formic acid—long term—and biotechnical method in summer followed by oxalic acid—trickling in winter—with 23.8% (95% CI: 15.0–35.6%) and the single treatment (O) formic acid—short term in summer—with 22.4% (95% CI: 15.2–31.9%). The widest confidence interval was observed for the combination (P) oxalic acid—trickling in summer and winter—with a loss rate of 24.3% (95% CI: 11.5–44.2%).

Thymol was only found in one of the generated combinations: (T) thymol in summer and oxalic acid—trickling in summer and winter—resulting in a loss rate of 20.1% (95% CI: 9.8–36.9%).

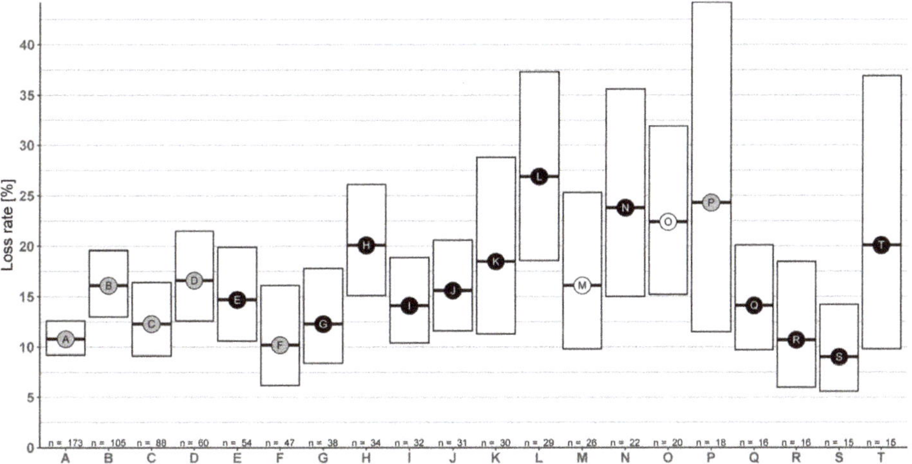

Figure 12. Winter honey bee colony loss rates (and 95% CI) resulting from combinations of various varroa control methods ordered by number of participants using the given combination: The point fill colour represents the number of different methods (single method = white, two methods = gray, and three methods = black). Please refer to Table 2 for explanation of the combination letter.

Table 2. This table shows the 20 most used combinations of treatment methods in the survey ($n > 14$), ordered by number of participants using the combination from highest to lowest for summer (June–October) and Winter (November–January).

Letter	Comb.	Methods					
A	2	SUMMER	Formic acid—long term	WINTER	Oxalic acid—trickling		
B	2	SUMMER	Formic acid—short term	WINTER	Oxalic acid—trickling		
C	2	SUMMER	Formic acid—long term	WINTER	Oxalic acid—sublimation		
D	2	SUMMER	Formic acid—short term	WINTER	Oxalic acid—sublimation		
E	3	SUMMER	Formic acid—short term	SUMMER	Formic acid—long term	WINTER	Oxalic acid—trickling
F	2	SUMMER	Oxalic acid—sublimation	WINTER	Oxalic acid—sublimation		
G	3	SUMMER	Formic acid—short term	SUMMER	Formic acid—long term	WINTER	Oxalic acid—sublimation
H	3	SUMMER	Formic acid—long term	SUMMER	Oxalic acid—sublimation	WINTER	Oxalic acid—sublimation
I	3	SUMMER	Formic acid—long term	SUMMER	Oxalic acid—trickling	WINTER	Oxalic acid—trickling
J	3	SUMMER	Formic acid—short term	SUMMER	Oxalic acid—sublimation	WINTER	Oxalic acid—sublimation
K	3	SUMMER	Biotechnical method	SUMMER	Oxalic acid—sublimation	WINTER	Oxalic acid—sublimation
L	3	SUMMER	Formic acid—short term	SUMMER	Oxalic acid—trickling	WINTER	Oxalic acid—trickling
M	1	SUMMER	Formic acid—long term				
N	3	SUMMER	Biotechnical method	SUMMER	Formic acid—long term	WINTER	Oxalic acid—trickling
O	1	SUMMER	Formic acid—short term				
P	2	SUMMER	Oxalic acid—trickling	WINTER	Oxalic acid—trickling		
Q	3	SUMMER	Biotechnical method	SUMMER	Oxalic acid—trickling	WINTER	Oxalic acid—trickling
R	3	SUMMER	Formic acid—short term	SUMMER	Oxalic acid—trickling	WINTER	Oxalic acid—sublimation
S	3	SUMMER	Biotechnical method	SUMMER	Formic acid—short term	WINTER	Oxalic acid—trickling
T	3	SUMMER	Oxalic acid—trickling	SUMMER	Thymol	WINTER	Oxalic acid—trickling

4. Discussion

4.1. Survey Data and Overall Losses

Compared to previous years, honey bee colony winter mortality in Austria 2018/2019 was at an average [26]. The individual loss rates in the survey are not normally distributed (Figure 1C), which is one of the reasons why a GZLM was used for the analysis [15]. More than half of the participants (69.7%) suffered no losses at all or a loss rate lower than 20%. Therefore, it should be feasible for the majority of beekeepers with lower losses to restock the lost colonies by themselves over the next summer season [13]. A multiple winter analysis revealed that, after winters with high losses, more new colonies are created over the summer season and low loss winters result in lower net gain of new colonies in the following season [26].

The variation of loss rates in different states does not seem to be a phenomenon restricted to Austria but can also be observed in the Czech Republic or the USA, where different regions show contrasting loss rates [6,26,27]. In other studies, this was explained by the fact that different regions consist of a divergent composition of landscape which could influence honey bee colony development and winter mortality [20,28]. In addition to this, weather effects have been found to influence colony survival [19]. Weather conditions also influence the hive management practices and varroa treatments examined in this article, but this lies beyond the scope of this study.

Different loss rates with respect to the elevation above sea level of the main winter apiary locations could be explained by more than one factor (Figure 3). One of those could be the climatic difference between elevation groups. In a previous study conducted in Austria, colder mean temperatures in September did result in lower overwintering loss rates [19]. Colder temperatures could infer longer winters and, therefore, a shorter breeding season, which results in lower varroa mite pressure. As mentioned before, the change in landscape could also play a role [20]. The latter study showed that semi-natural areas, pastures, and coniferous forests had a positive effect on colony survival in Austria. The lower loss rates with wintering colonies above 600 m elevation could further be explained by fewer honey bee colonies in higher regions and, connected to this, lower spread of viruses or other bee pathogens [29–31].

4.2. Queen Management

Queen problems after winter and winter colony loss rate are influenced by many factors, such as biological causes or beekeeping management practices [2,32,33]. Compared to previous years, the special case of winter colony losses due to unsolvable queen problems (living colonies without a laying queen or a drone laying queen) seem stable within 3.6–4.4% [26] and are similar to many other countries participating in COLOSS surveys, with 4–5% [17]. Only 7.8% of the participants experienced more queen problems in their colonies during the 2018 foraging season compared to what they usually observed (Figure 5A). More queen-related problems seemed to go hand in hand with a higher queen-related loss rate after winter (Figure 5B). A study from the USA identified "queen events", i.e., colonies with emergency or supersedure queen cells, as a significant negative factor for colony survival [34]. Possible causes for more queen problems could be led back to neonicotinoids [35] or even package transport of queen bees [36]. These findings and the given results underline the importance of queen bees to colony success [33].

One of the biological influences is likely the age of queen bees going into winter. Several studies found that old queens lower the chance of colony survival [11,12,18,37]. Interestingly, we found significantly lower queen-related losses and winter colony losses (excluding queen-related losses) when the participants exchanged more than one-fourth of their old queens with younger ones in the season before winter (Figure 4C,D). This is in accordance with References [11,18], where each percentage of new queens resulted in a small increase of colony survival. Thus, we conclude that a healthy young and well-mated queen is an important factor for overwintering survival and colony health, probably because a younger queen can build a stronger colony than an older queen [33,38].

Therefore, to replace old queens each year seems to be practical and could lower the winter loss rate [11,37].

4.3. Hive Management Practices

Beekeeping management practices and operational factors are directly influenced by the individual beekeeper [39]. In international analyses of COLOSS surveys, large sample sizes are used to identify beekeeping practices that reduce colony mortality [7]. The power of big data sets (with varying participation from different countries) may obscure regionally important results. We therefore use the opportunity to investigate the efficacy and importance of hive management practices for Austria.

We found a lower winter colony loss rate for migratory beekeeping operations compared to non-migrating operations (Figure 6B). The reasons for this could be that migratory beekeepers are more experienced and that migrated colonies have access to better foraging sources [18]. In Austria, colonies are mainly migrated to harvest special honeys, while migrating colonies for paid pollination service is rarely utilized. However, this effect was not consistently found in Austria for some previous investigations [13,40]. In a multi-country analysis of the winter season of 2016/2017 [16], most countries, including Austria, did show no significant difference between migrating and non-migrating beekeepers; however, in the following year, an effect could be found [17]. US beekeepers that migrated their colonies into almond fields experienced a higher total loss rate for the winter 2007/2008 [6]. In contrast to this, two years later, the results indicated significantly lower total losses [27].

Due to the fact that most beekeepers in Austria are hobbyists or sideline beekeepers, the operation size of most participants is relatively small [13]. Hence, most survey participants own a small amount of the total colonies whereas a small number of participants own a significant amount of colonies in Austria (Figure 1A,B). Significant different loss rates among operation sizes were already demonstrated before [12,13,17,41]. Beekeepers with a smaller number of colonies are at a greater risk of losing their colonies (Figure 8). Beekeepers who manage more colonies are likely to have better training and more experience than hobbyist beekeepers, a hypothesis supported by another Austrian study [12].

Over the years of conducting this survey, beekeepers requested to further investigate effects of some operational factors, which they thought might mitigate colony losses. Of these operational factors, only one could be statistically verified, underlining the importance of wax for honey bee health. If participants did purchase wax from outside their own operation, they had a higher loss rate than beekeepers using only their own wax. Bees are exposed to various pesticides, and beeswax, due to its chemical character, is the most contaminated beehive matrix and a bio accumulator of acaricides, fungicides, and insecticides over years [42]. Residues in the wax are frequently found in multiple countries [43,44]. A study in which beeswax foundations were artificially contaminated with pesticide resulted in no negative effect on colony survival [45] but could still influence the survival of bees in addition with other factors. Viruses and spores have also be found in beeswax, and it can therefore pose as a possible viral reservoir [46]. In the cited study, the removal of viral pathogens from old frames did not make a significant difference in the probability of colony survival, and varroa and its varroa-transmitted viruses were considered the greater problem [46]. However, removing pathogens still lowered the chance of infected emerging broods. Commercial wax producers do commonly heat the wax to eliminate American foulbrood spores, which also can remove other pathogens, but this has no effect on pesticides [44]. Thus, the reason why participants who purchased wax from outside their own operation had a higher loss rate is not entirely clear to us. It is possible that participants who bought wax from outside their own operation are beginners who do not possess the equipment or resources for executing their own wax cycle. The result of our study suggests that the factor "buying-in wax" may be a proxy for a certain management or degree of professionalism. We suggest further investigation of wax quality and origin as a risk factor for honey bee colony losses. Overall, colonies with new combs are healthier than colonies with old combs [47,48]. Nevertheless, the sourcing and previous treatment of foreign wax could be an important factor to lowering the amount of residue in colonies near agriculture fields.

We found no difference in winter loss rate between colonies on natural comb or on a foundation of wax (Figure 6H). Replacing old brood frames had no significant influence on colony survival but showed a tendency for lower loss rates with exchange rates above 30% (Figure 7). Comparing this year to the previous years in Austria (2013/14, 2014/15, 2015/16, and 2016/17) two of the years showed a significantly lower loss rate while the other two did not [40]. One possible explanation for this fact could be the varroa treatment strategy "other biotechnical methods" and the lower loss rate associated with this method (Figure A3C), which often includes removal of old brood frames as a side effect.

Certificated organic beekeeping or nonorganic operations had no different probability of honey bee colony winter loss (Figure 6A). The European organic regulation (EC No. 834/2007, 889/2008) is the minimum standard for other organic authorities in the EU. The main restrictions for organic certified beekeepers are the mandatory use of organic certificated sugar/syrup for feeding, the prohibition for the use of synthetic treatments against mites or other pests, and the compulsory use of comb wax from organic beekeeping operations [49]. The location of organic certified apiaries could influence the winter loss rate but is not well defined in the organic regulation and can be differently interpreted by organic control bodies; the same can be said for wax which should be free of contamination by substances not authorized for organic production [49]. As the number of survey participants using synthetic treatments was quite low (Figure 10L) [26], we conclude that feeding and wax quality are the only major differences between conventional and organic beekeeping in our study.

The type of hive (hives fabricated from synthetic materials, insulated hives, or open screened bottom board in winter) had no negative or positive influence on the loss rate (Figure 6D–F). Previous analysis of these factors in Austria came to the same conclusion [40]. A study in Spain examined the temperature and humidity in hives with open screened bottom boards and without, but there seems to be no crucial difference between the hive types regarding colony health [50]. Naturally, the lowest outside temperature in this experiment was around 7 °C and is not comparable to cold winter periods in Austria.

Breeding lines with queens bred from varroa tolerant/resistant stocks had no different colony mortality than others (Figure 6C). Such breeding lines are often selected based on the amount of removed damaged brood, for example, via freeze-killed brood assay. In field studies from the US, such lines showed reduced mites in worker brood and adult bees [51]. Beekeepers from the US with varroa resistant bee stocks also experienced lower loss rates than those without [52]. It is probably difficult for the individual beekeeper to check if most of their honey bees have traits favourable for survival, but it shows that more research on this topic is needed.

It should be discussed for future surveys if beekeeping management questions, which obviously do not influence the colony survival over winter in multiple years and countries, should be removed to minimize the amount of time spent by the participants to conclude the survey. However, these factors could still have an effect in combination with other factors. On the other hand, most of the questions discussed in this section resulted from participatory processes and reflect the interest of beekeepers.

4.4. Forage as Risk Factor

Colonies reported to be foraging on maize showed significantly higher winter loss rates compared to colonies that did not (Figure A1B). Maize does not produce nectar but pollen, which can be directly collected by bees; however, it is not a preferred source for honeybees [53,54]. Still, some cases of colonies collecting large amounts of maize pollen have been documented in Austria [55]. A possible explanation for the higher loss rate could be pesticide contaminated pollen, residues in guttation water, or other indirect ways of getting into contact with agriculture chemicals used in maize fields [18,54,56]. Other reasons for our result include poor landscape for foraging with a lot of maize fields nearby, which presumably increases honey bees collecting poor-quality maize pollen or guttation water as well as the lack of nutritive pollen [18,57]. Maize pollen foraging is difficult to assert by participants, but this causality is not needed to explain our results, as this may only describe the quality of maize growing landscapes versus environments without maize. This was also proposed as potential risk in the multi-country analysis from Gray et al. [17].

Late catch crop likewise caused a higher probability of loss (Figure A1D), which at least was not observed in winter 2017/2018 for Austria [17]. Possible reasons could be an extended brood period due to late honey and pollen flows or, again, contact with pesticides in these fields [17].

Honeydew had no influence on colony survival over winter, but participants experiencing a melezitose forage, which often comes in areas with honeydew, had significantly higher loss rates (Figure A1E,F). Melezitose sometimes appears in July or in autumn and has been a familiar problem for beekeepers for a long time. Melezitose fills the brood and honey frames with hard to remove crystalline honey, and bees can invert only a small percentage of the collected melezitose sugar in comparison to sucrose [58]. Furthermore, colonies overwintering on it are often affected of dysentery, which is attributed to the high mineral content [59]. The time when melezitose forage occurs could play a crucial role and is not the same each year. Though melezitose honey is poorly studied, it is a well-recognized problem for beekeepers located in dedicated areas of Austria.

Sunflower and oilseed rape forages are often discussed as risk factors for honey bees. In this study, bees foraging on sunflower could not be linked to raised colony losses (Figure A1C), although this was found in previous years [17,40]. Observed oilseed rape foraging could also not be associated with higher loss rates (Figure A1A). Contrary results are published for Austria and other European countries in winter 2013/2014 [18] and 2017/2018 [17], though some countries experienced the opposite. Oilseed rape provides abundant nectar and pollen for bees, but insecticidal treatments might affect colony development and survival of bees [60–62]. Finally, similar as discussed for maize, diversity of forage due to monoculture fields in such areas could be low [63].

4.5. Varroa Control

Varroa destructor is regarded as the greatest threat to apiculture. Beekeepers need to efficiently treat their colonies or they might face their collapse within 3–4 years [10,12]. The majority of participants stated to monitor varroa infestation levels. Those beekeepers experienced a significantly lower loss rate (Figure 9A). Though this monitoring alone does not decrease varroa levels, such practices can be promoted as good beekeeping practices.

Observations of bees with crippled/deformed wings were associated with a higher loss rate (Figure 9B). Crippled bees can emerge due to cold or viral damage, but they are most often connected to the mite transmitted deformed wing virus (DWV). Therefore, if beekeepers see such bees, this can be interpreted as an alarm signal and counter measures must immediately be taken. The need to monitor DWV load and varroa mite infestation was demonstrated in southern Spain, where high DWV load and high varroa counts resulted in weaker colonies and a higher probability of losing colonies but did not show a significant correlation between DWV symptoms and viral load [64]. This supports the notion that mite-related damage strongly influences the winter loss rate and is one of the crucial factors for colony losses in Austria [12].

The spectrum of varroa control methods applied in Austria is rather limited compared to other countries [26]. To treat colonies, most participants follow the recommendation from the Austrian Agency for Health and Food Safety (AGES) [65] to evaporate formic acid after honey harvest and trickle or sublimate oxalic acid products in winter (Table 2C–F). The benefits of organic acids are a low risk of resistance, a low risk of residues, and a good efficacy against *V. destructor* [10]. Beekeepers in Austria do not commonly apply synthetic acaricides (Figure 10L). Therefore, we pooled the few applications of different types of synthetic acaricides (e.g., Amitraz, Coumaphos, etc.). These agents could lead to wax residues or pollution of honey [10]. The residues in wax could account for the various complications in the bee brood stage [45]. Nevertheless, the synthetic acaricide Amitraz did result in lower loss rates than other varrocide products (synthetic and organic) in a study from the US [52].

In spring (April–May), the most common control method in Austria was drone brood removal (Figure A2A). Participants who performed this method had a significantly lower loss rate the upcoming winter period. If drone brood removal was done in spring and summer (June–October), there is a trend to lose less colonies compared to only removing drone brood in spring or summer (Figure 11). This result was also observed in Austria in other years [40]. In a field study from the US, frequent removal of drone brood resulted in lower mite infestation [66]. However, in comparison to participants who did not remove drone brood at all, there is no significant difference in loss rates. More research on this method under local environmental conditions needs to be done to possibly enhance the positive effect.

In summer (June–October). significantly lower loss rates were observed for participants applying "other biotechnical methods", e.g., trapping comb or complete brood removal to control mites (Figure A3C). These effective methods are often labour intensive [10]. In spring, biotechnical methods (excluding drone brood removal or hyperthermia) are not often used in Austria but could be considered as good practice to fight *V. destructor* with rising temperatures and a prolonged brood period in the future. This was shown in a study in Italy (Reggio Emilia, Po Valley) where the caging of the queen in spring produced no negative impact on honey harvest or brood amount but resulted in a lower mite infestation rate [67]. This could also encourage beekeepers aiming for late honey flows, which they would otherwise miss because the mite population is already too high. Participants applying oxalic acid by trickling in summer experienced high winter losses (Figure A3H). Possible reasons for this could be multiple tricklings and a negative effect on bee health or remaining broods in the colonies which leads to insufficient treatment success [68]. Therefore, oxalic acid trickling is recommended to be performed only once in the broodless period [10,69].

In winter (November–January), oxalic acid represents the dominant choice of treatment with no differences between application by trickling or sublimation (Figure A3A,B). Both treatments have already been evaluated to be very effective in the broodless period, and there is currently no alternative available [10,69].

So far, we solely discussed single treatment methods, but integrated varroa control strategies are comprised of combinations of different treatments. We therefore identified the most common combinations of treatments (Table 2, Figure 12). This allows, for example, further examination of the shown negative effect of oxalic acid—trickling in summer—in combination with other treatments. We found high loss rates for the combination (L) formic acid—short term in summer—and oxalic acid—trickling in summer and winter (Table 2L) and the combination (P) only oxalic acid—trickling in summer and winter (Table 2P). We conclude that trickling in summer is either not effective or causes negative effects on bee health. In combination with formic acid—long term in summer (Table 2I), this resulted in average losses. This might be due to the positive impact of formic acid—long-term evaporation.

Two frequently applied combinations were long-term evaporation of formic acid in summer and oxalic acid trickling or sublimation in winter (Table 2A,C), which both resulted in average colony losses. Formic acid is the only allowed organic acid in Austria which is effective against phoretic and reproductive mites [10]. Our results hence support the current recommendation from AGES to use formic acid in summer and oxalic acid in winter [65]. Long-term evaporation should be preferred, as short term may not be as efficient in mite reduction. Short-term evaporation in combination with oxalic acid in winter (Figure 12B,D) and, additionally, the double application of formic acid—long and short term in summer—with oxalic acid—trickling in winter (Figure 12E)—had higher loss rates than the single application of formic acid—long term in summer—and oxalic acid—trickling in winter (Figure 12A). We assume that either the application of both formic acid variants could lead to brood damage or the double application is an emergency measure because of high varroa counts.

The combination of biotechnical methods and short-term formic acid evaporation in summer plus oxalic acid—trickling in winter (Table 2S)—showed one of the lowest loss rates. Care must be taken in interpreting these results, as only 15 participants exercised this combination. Further, the biotechnical methods asked for in our survey could include a wide range of various procedures. To learn more about efficacy of different biotechnical methods, further studies should specify which biotechnical methods are used. Though this would lower sample size, discrepant results with biotechnical methods in combination with other methods could be better understood.

The combination of oxalic acid—sublimation in summer and winter (Table 2F) resulted in low loss rates. This method seems to be efficient to treat mite infestation and sublimation in comparison to trickling in summer is favourable. The sublimation of oxalic acid does not reduce reproductive mites [10,69]. Therefore, the colony must be broodless or sublimation is repeated multiple times in summer. This combination could offer a potential to represent a reliable method but would require a more in depth analysis with field studies on the amount of oxalic acid, on frequency of application, and the different sublimation equipment.

There was only one frequent combination with thymol. Preparations with the essential oil (in summer) and oxalic acid—trickling in summer and winter (Table 2T)—resulted in high loss rates but with a wide confidence interval. In an Australian study on thymol and the beneficial effect on hygienic behaviour, researchers found inconsistent results [70]. They proposed that different factors play a role for this differential outcome, such as environmental or genetic differences. In the loss rates presented here for Austria, thymol was combined with oxalic acid—trickling in summer. As we already discussed that oxalic acid trickling in summer should not be performed, we cannot recommend this treatment combination. This is no general recommendation against thymol usage, as we are lacking reliable data. In this study, we identified some unexpected combinations of varroa control methods applied in Austria. We recommend further research to better understand the motivation of beekeepers behind those and direct them to reasonable and effective treatment plans.

5. Conclusions

The winter 2018/2019 represented an average year in terms of overwinter honey bee colony losses in Austria. Significantly different loss rates were observed among the states, which could be led back to a divergent composition of landscapes or regional differences in weather and varroa pressure [19]. Low colony losses were associated with indicators for advanced beekeeping: migratory beekeeping or keeping a larger number of colonies. The majority of operational decisions that beekeepers can make in Austria, like certified organic beekeeping or having insulated hives in winter, had no effect on winter mortality. The only operational decision leading to lower losses was the self-sustaining operation of wax cycles, again a sign of a certain degree of professionalism. Queen problems during the summer resulted in a higher queen-related loss rate. Young queens showed a beneficial effect on both colony survival and queen-related losses. The observation of a notable number of bees with deformed wings in summer can be interpreted as an alarm signal for beekeepers, as this resulted in high colony losses. Access to various crops or honey flows was confirmed to impact colony survival—the exact drivers behind these need further investigation [17]. Varroa treatment with biotechnical methods in summer had a favourable effect on winter survival. For the first time, we investigated different commonly applied combinations of varroa control methods on winter colony losses. The combination of long-term evaporation of formic acid in summer and oxalic acid usage in winter, the dominant and officially recommended varroa control strategy in Austria, resulted in an average loss rates. The lowest loss rates were observed for biotechnical methods and short-term evaporation of formic acid in summer followed by oxalic acid trickling in winter. The best treatment combination without formic acid was oxalic acid sublimation in summer and winter. It still needs to be considered that the *V. destructor* population cycle is very complex and that treatment methods do not return the anticipated results each year, which could be led back to weather conditions or other environmental factors. Therefore, a multiple year analysis on treatment combinations is required to provide more reliable insights.

Our study supports how diverse the factors influencing honey bee colony mortality can be and how difficult it is to determine one factor alone to improve colony survival. It is substantial to repeat these studies and to make multi-year and multi-country analyses in order to maintain relevancy in light of current trends in beekeeping practices and climate change. Our goal is to assist beekeepers with statistical analysis and to improve beekeeping and the honey bee population in Austria with empirically gained knowledge. One measure, based on this study, is to increase professionalism in beekeeping to reduce winter colony losses.

Author Contributions: Conceptualization, R.B.; methodology, R.B.; software, H.O.; validation, H.O.; formal analysis, H.O.; investigation, R.B. and H.O.; resources, R.B.; data curation, H.O.; writing—original draft preparation, H.O.; writing—review and editing, H.O. and R.B.; visualization, H.O.; supervision, R.B.; project administration, R.B. All authors have read and agreed to the published version of the manuscript.

Funding: This research was funded by ZUKUNFT BIENE 2 grant number 101295/2.

Acknowledgments: We want to thank all beekeepers that participated in our study or helped with discussions or suggestions to improve the study aims. We also want to thank the monitoring core project of the nonprofit honey bee research association COLOSS (www.coloss.org).

Conflicts of Interest: The authors declare no conflict of interest.

Abbreviations

The following abbreviations are used in this manuscript:

AGES	the Austrian Agency for Health and Food Safety
CI	confidence interval
COLOSS	prevention of honey bee COlony LOSSes
DWV	deformed wing virus
GZLM	generalized linear model

Appendix A

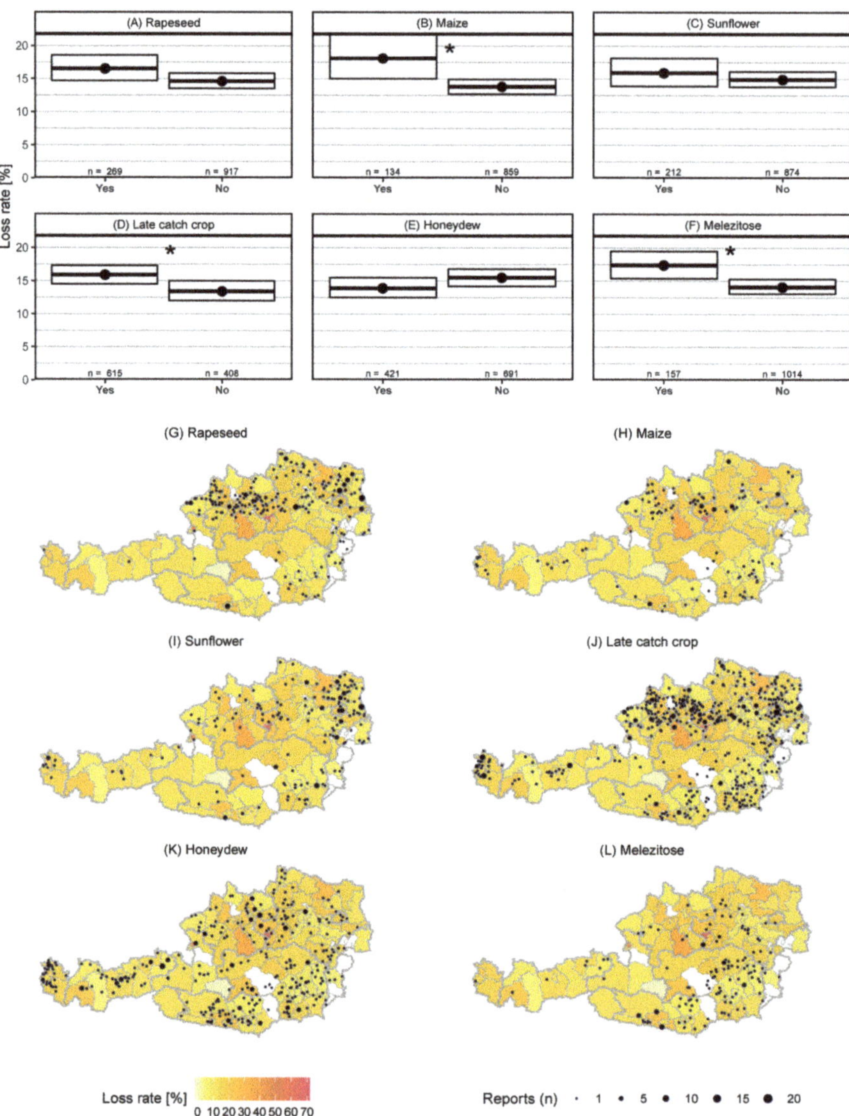

Figure A1. Six different types of crops (A = oilseed rape—*Brassica napus*, B = maize—*Zea mays*, C = sunflower—*Helianthus annuus*, D = late catch crop, E = honeydew, F = Melezitose) and their winter honey bee colony loss rates (and 95% CI): The maps indicate the rough location of the main wintering apiary of the participants. Multiple crop reports were possible, which means that different crops could apply to the same beekeeper.

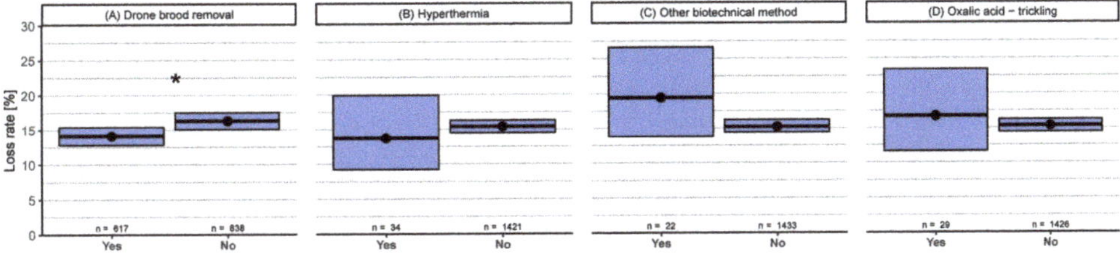

Figure A2. Treatment methods (yes = *n* > 19) used in spring 2018 (April–May) as single factors to winter honey bee colony loss rates (and 95% CI): To improve sample size, the oxalic mixture was added to oxalic trickling.

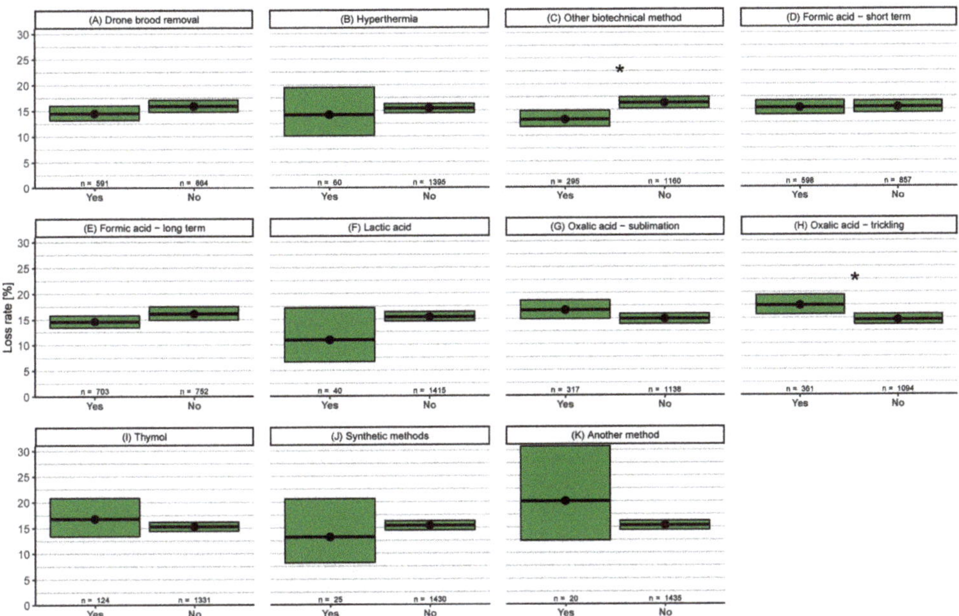

Figure A3. Treatment methods (yes = *n* > 19) used in summer 2018 (June–October) as single factors to winter honey bee colony loss rates (and 95% CI): To improve sample size, synthetic methods were grouped together and the oxalic mixture was added to oxalic trickling.

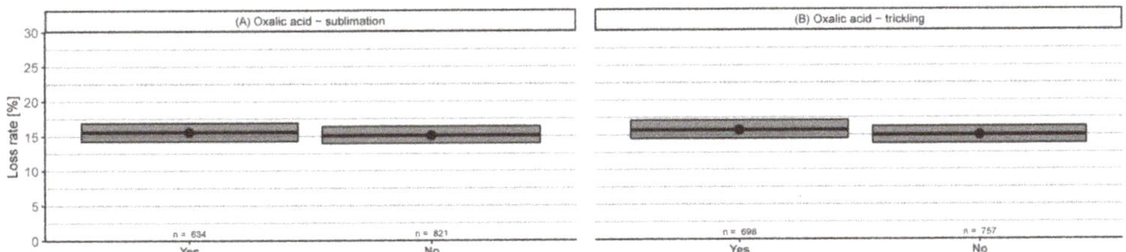

Figure A4. Treatment methods (yes = *n* > 19) used in autumn/winter 2018 (November–January) as single factors to winter honey bee colony loss rates (and 95% CI): To improve sample size, the oxalic mixture was added to oxalic trickling.

References

1. Gallai, N.; Salles, J.M.; Settele, J.; Vaissière, B.E. Economic valuation of the vulnerability of world agriculture confronted with pollinator decline. *Ecol. Econ.* **2009**, *68*, 810–821. [CrossRef]
2. Steinhauer, N.; Kulhanek, K.; Antúnez, K.; Human, H.; Chantawannakul, P.; Chauzat, M.P.; van Engelsdorp, D. Drivers of colony losses. *Curr. Opin. Insect Sci.* **2018**, *26*, 142–148. [CrossRef]
3. Belsky, J. Impact of Biotic and Abiotic Stressors on Managed and Feral Bees. *Insects* **2019**, *10*, 233. [CrossRef]
4. Neov, B.; Georgieva, A.; Shumkova, R.; Radoslavov, G.; Hristov, P. Biotic and Abiotic Factors Associated with Colonies Mortalities of Managed Honey Bee (Apis mellifera). *Diversity* **2019**, *11*, 237. [CrossRef]
5. Brodschneider, R.; Crailsheim, K. Nutrition and health in honey bees. *Apidologie* **2010**, *41*, 278–294. [CrossRef]
6. Van Engelsdorp, D.; Hayes, J.; Underwood, R.M.; Pettis, J. A survey of honey bee colony losses in the U.S., Fall 2007 to Spring 2008. *PLoS ONE* **2008**, *3*, e4071. [CrossRef]
7. Van der Zee, R.; Pisa, L.; Andonov, S.; Brodschneider, R.; Charrière, J.D.; Chlebo, R.; Coffey, M.F.; Crailsheim, K.; Dahle, B.; Gajda, A.; et al. Managed honey bee colony losses in Canada, China, Europe, Israel and Turkey, for the winters of 2008–9 and 2009–10. *J. Apic. Res.* **2012**, *51*, 100–114. [CrossRef]
8. Thoms, C.A.; Nelson, K.C.; Kubas, A.; Steinhauer, N.; Wilson, M.E.; van Engelsdorp, D. Beekeeper stewardship, colony loss, and Varroa destructor management. *Ambio* **2019**, *48*, 1209–1218. [CrossRef]
9. Underwood, R.M.; Traver, B.E.; López-Uribe, M.M. Beekeeping Management Practices Are Associated with Operation Size and Beekeepers' Philosophy towards in-Hive Chemicals. *Insects* **2019**, *10*, 10. [CrossRef]
10. Rosenkranz, P.; Aumeier, P.; Ziegelmann, B. Biology and control of Varroa destructor. *J. Invertebr. Pathol.* **2010**, *103*, S96–S119. [CrossRef]
11. Genersch, E.; von der Ohe, W.; Kaatz, H.; Schroeder, A.; Otten, C.; Büchler, R.; Berg, S.; Ritter, W.; Mühlen, W.; Gisder, S.; et al. The German bee monitoring project: A long term study to understand periodically high winter losses of honey bee colonies. *Apidologie* **2010**, *41*, 332–352. [CrossRef]
12. Morawetz, L.; Köglberger, H.; Griesbacher, A.; Derakhshifar, I.; Crailsheim, K.; Brodschneider, R.; Moosbeckhofer, R. Health status of honey bee colonies (Apis mellifera) and disease-related risk factors for colony losses in Austria. *PLoS ONE* **2019**, *14*, e0219293. [CrossRef] [PubMed]
13. Brodschneider, R.; Moosbeckhofer, R.; Crailsheim, K. Surveys as a tool to record winter losses of honey bee colonies: A two year case study in Austria and South Tyrol. *J. Apic. Res.* **2010**, *49*, 23–30. [CrossRef]
14. Neumann, P.; Carreck, N.L. Honey bee colony losses. *J. Apic. Res.* **2009**, *49*, 1–6. [CrossRef]
15. Van der Zee, R.; Gray, A.; Holzmann, C.; Pisa, L.; Brodschneider, R.; Chlebo, R.; Coffey, M.F.; Kence, A.; Kristiansen, P.; Mutinelli, F.; et al. Standard survey methods for estimating colony losses and explanatory risk factors in Apis mellifera. *J. Apic. Res.* **2013**, *52*, 1–36. [CrossRef]
16. Brodschneider, R.; Gray, A.; Adjlane, N.; Ballis, A.; Brusbardis, V.; Charrière, J.D.; Chlebo, R.; Coffey, M.F.; Dahle, B.; de Graaf, D.C.; et al. Multi-country loss rates of honey bee colonies during winter 2016/2017 from the COLOSS survey. *J. Apic. Res.* **2018**, *57*, 452–457. [CrossRef]
17. Gray, A.; Brodschneider, R.; Adjlane, N.; Ballis, A.; Brusbardis, V.; Charrière, J.D.; Chlebo, R.; Coffey, M.F.; Cornelissen, B.; Amaro da Costa, C.; et al. Loss rates of honey bee colonies during winter 2017/18 in 36 countries participating in the COLOSS survey, including effects of forage sources. *J. Apic. Res.* **2019**, 1–7. [CrossRef]
18. Van der Zee, R.; Brodschneider, R.; Brusbardis, V.; Charrière, J.D.; Chlebo, R.; Coffey, M.F.; Dahle, B.; Drazic, M.M.; Kauko, L.; Kretavicius, J.; et al. Results of international standardised beekeeper surveys of colony losses for winter 2012/2013: Analysis of winter loss rates and mixed effects modelling of risk factors for winter loss. *J. Apic. Res.* **2014**, *53*, 19–34. [CrossRef]
19. Switanek, M.; Crailsheim, K.; Truhetz, H.; Brodschneider, R. Modelling seasonal effects of temperature and precipitation on honey bee winter mortality in a temperate climate. *Sci. Total Environ.* **2017**, *579*, 1581–1587. [CrossRef]
20. Kuchling, S.; Kopacka, I.; Kalcher-Sommersguter, E.; Schwarz, M.; Crailsheim, K.; Brodschneider, R. Investigating the role of landscape composition on honey bee colony winter mortality: A long-term analysis. *Sci. Rep.* **2018**, *8*, 12263. [CrossRef]
21. GeoNames. GeoNames—Geographical Database. Available online: https://www.geonames.org/ (accessed on 1 September 2019).

22. R Core Team. *R: A Language and Environment for Statistical Computing*; R Foundation for Statistical Computing: Vienna, Austria, 2018.

23. Wickham, H. *Ggplot2: Elegant Graphics for Data Analysis*; Use R!; Springer: New York, NY, USA, 2009; OCLC: ocn382399721.

24. Oberreiter, H. *Codebase: Austria Honey Bee Colony Overwinter Losses 2018/2019*; Zenodo: Genève, Switzerland, 2019. [CrossRef]

25. Nijenhuis, A.; Wilf, H. *Combinatorial Algorithms for Computers and Calculators*; Academic Press: New York, NY, USA, 1978.

26. Brodschneider, R.; Brus, J.; Danihlík, J. Comparison of apiculture and winter mortality of honey bee colonies (*Apis mellifera*) in Austria and Czechia. *Agric. Ecosyst. Environ.* **2019**, *274*, 24–32. [CrossRef]

27. VanEngelsdorp, D.; Hayes, J.; Underwood, R.M.; Caron, D.; Pettis, J. A survey of managed honey bee colony losses in the USA, fall 2009 to winter 2010. *J. Apic. Res.* **2010**, *50*, 1–10. [CrossRef]

28. Van Esch, L.; De Kok, J.L.; Janssen, L.; Buelens, B.; De Smet, L.; de Graaf, D.C.; Engelen, G. Multivariate Landscape Analysis of Honey Bee Winter Mortality in Wallonia, Belgium. *Environ. Model. Assess.* **2019**. [CrossRef]

29. Seeley, T.D.; Smith, M.L. Crowding honeybee colonies in apiaries can increase their vulnerability to the deadly ectoparasite Varroa destructor. *Apidologie* **2015**, *46*, 716–727. [CrossRef]

30. Forfert, N.; Natsopoulou, M.E.; Paxton, R.J.; Moritz, R.F. Viral prevalence increases with regional colony abundance in honey bee drones (*Apis mellifera* L). *Infect. Genet. Evol.* **2016**, *44*, 549–554. [CrossRef]

31. Dynes, T.L.; Berry, J.A.; Delaplane, K.S.; Brosi, B.J.; de Roode, J.C. Reduced density and visually complex apiaries reduce parasite load and promote honey production and overwintering survival in honey bees. *PLoS ONE* **2019**, *14*, e0216286. [CrossRef]

32. Döke, M.A.; Frazier, M.; Grozinger, C.M. Overwintering honey bees: Biology and management. *Curr. Opin. Insect Sci.* **2015**, *10*, 185–193. [CrossRef]

33. Amiri, E.; Strand, M.K.; Rueppell, O.; Tarpy, D.R. Queen quality and the impact of honey bee diseases on queen health: Potential for interactions between two major threats to colony health. *Insects* **2017**, *8*, 22–26. [CrossRef]

34. VanEngelsdorp, D.; Tarpy, D.R.; Lengerich, E.J.; Pettis, J.S. Idiopathic brood disease syndrome and queen events as precursors of colony mortality in migratory beekeeping operations in the eastern United States. *Prev. Vet. Med.* **2013**, *108*, 225–233. [CrossRef]

35. Williams, G.R.; Troxler, A.; Retschnig, G.; Roth, K.; Yañez, O.; Shutler, D.; Neumann, P.; Gauthier, L. Neonicotinoid pesticides severely affect honey bee queens. *Sci. Rep.* **2015**, *5*, 14621. [CrossRef]

36. Withrow, J.M.; Pettis, J.S.; Tarpy, D.R. Effects of Temperature During Package Transportation on Queen Establishment and Survival in Honey Bees (Hymenoptera: Apidae). *J. Econ. Entomol.* **2019**, *112*, 1043–1049. [CrossRef] [PubMed]

37. Giacobino, A.; Molineri, A.; Cagnolo, N.B.; Merke, J.; Orellano, E.; Bertozzi, E.; Masciangelo, G.; Pietronave, H.; Pacini, A.; Salto, C.; et al. Queen replacement: The key to prevent winter colony losses in Argentina. *J. Apic. Res.* **2016**, *55*, 335–341. [CrossRef]

38. Ricigliano, V.A.; Mott, B.M.; Floyd, A.S.; Copeland, D.C.; Carroll, M.J.; Anderson, K.E. Honey bees overwintering in a southern climate: Longitudinal effects of nutrition and queen age on colony-level molecular physiology and performance. *Sci. Rep.* **2018**, *8*, 1–11. [CrossRef] [PubMed]

39. Jacques, A.; Laurent, M.; Consortium, E.; Ribière-Chabert, M.; Saussac, M.; Bougeard, S.; Budge, G.E.; Hendrikx, P.; Chauzat, M.P. A pan-European epidemiological study reveals honey bee colony survival depends on beekeeper education and disease control. *PLoS ONE* **2017**, *12*, e0172591. [CrossRef]

40. Crailsheim, K.; Moosbeckhofer, R.; Brodschneider, R. Future of Honey Bees—Basic Research for Project for Honey Bee Health and Bee Protection; Final Report Poject 'Zukunft Biene' 2014–2018 Austria; 2018. Available online: https://www.ages.at/en/topics/environment/bees/research-projects-on-bees/future-of-honey-bees/ (acessed on 7 February 2020)

41. Castilhos, D.; Bergamo, G.C.; Gramacho, K.P.; Gonçalves, L.S. Bee colony losses in Brazil: A 5-year online survey. *Apidologie* **2019**, *50*, 263–272. [CrossRef]

42. Calatayud-Vernich, P.; VanEngelsdorp, D.; Picó, Y. Beeswax cleaning by solvent extraction of pesticides. *MethodsX* **2019**, *6*, 980–985. [CrossRef]

43. Harriet, J.; Campá, J.P.; Grajales, M.; Lhéritier, C.; Gómez Pajuelo, A.; Mendoza-Spina, Y.; Carrasco-Letelier, L. Agricultural pesticides and veterinary substances in Uruguayan beeswax. *Chemosphere* **2017**, *177*, 77–83. [CrossRef]

44. Calatayud-Vernich, P.; Calatayud, F.; Simó, E.; Picó, Y. Pesticide residues in honey bees, pollen and beeswax: Assessing beehive exposure. *Environ. Pollut.* **2018**, *241*, 106–114. [CrossRef]

45. Payne, A.N.; Walsh, E.M.; Rangel, J. Initial Exposure of Wax Foundation to Agrochemicals Causes Negligible Effects on the Growth and Winter Survival of Incipient Honey Bee (*Apis mellifera*) Colonies. *Insects* **2019**, *10*, 19. [CrossRef]

46. De Guzman, L.I.; Simone-Finstrom, M.; Frake, A.M.; Tokarz, P. Comb Irradiation Has Limited, Interactive Effects on Colony Performance or Pathogens in Bees, *Varroa destructor* and Wax Based on Two Honey Bee Stocks. *Insects* **2019**, *10*, 15. [CrossRef]

47. Koenig, J.P.; Boush, G.M.; Erickson, E.H. Effect of type of brood comb on Chalk brood disease in honeybee colonies. *J. Apic. Res.* **1986**, *25*, 58–62. [CrossRef]

48. Berry, J.A.; Delaplane, K.S. Effects of comb age on honey bee colony growth and brood survivorship. *J. Apic. Res.* **2001**, *40*, 3–8. [CrossRef]

49. Thrasyvoulou, A.; Broeker, U.; Chrysoula, T.; Vilas-Boas, M.; Wallner, K.; Amsler, T.; Garces, S.; Lodesani, M.; Siceanu, A.; Westerhoff, A.; et al. Improvements To The Regulations On Organic Farming To Facilitate The Practice Of Organic Beekeeping. *Bee World* **2015**, *91*, 58–61. [CrossRef]

50. Sánchez, V.; Gil, S.; Flores, J.M.; Quiles, F.J.; Ortiz, M.A.; Luna, J.J. Implementation of an electronic system to monitor the thermoregulatory capacity of honeybee colonies in hives with open-screened bottom boards. *Comput. Electron. Agric.* **2015**, *119*, 209–216. [CrossRef]

51. Ibrahim, A.S.; Reuter, G.; Spivak, M. Field trial of honey bee colonies bred for mechanisms of resistance against Varroa destructor. *Apidologie* **2007**, *38*, 67–76. [CrossRef]

52. Haber, A.I.; Steinhauer, N.A.; vanEngelsdorp, D. Use of Chemical and Nonchemical Methods for the Control of Varroa destructor (Acari: Varroidae) and Associated Winter Colony Losses in U.S. Beekeeping Operations. *J. Econ. Entomol.* **2019**, *112*, 1509–1525. [CrossRef]

53. Höcherl, N.; Siede, R.; Illies, I.; Gätschenberger, H.; Tautz, J. Evaluation of the nutritive value of maize for honey bees. *J. Insect Physiol.* **2012**, *58*, 278–285. [CrossRef]

54. Urbanowicz, C.; Baert, N.; Bluher, S.E.; Böröczky, K.; Ramos, M.; McArt, S.H. Low maize pollen collection and low pesticide risk to honey bees in heterogeneous agricultural landscapes. *Apidologie* **2019**, *2011*. [CrossRef]

55. Brodschneider, R.; Gratzer, K.; Kalcher-Sommersguter, E.; Heigl, H.; Auer, W.; Moosbeckhofer, R.; Crailsheim, K. A citizen science supported study on seasonal diversity and monoflorality of pollen collected by honey bees in Austria. *Sci. Rep.* **2019**, *9*, 16633. [CrossRef]

56. Schmolke, A.; Kearns, B.; O'Neill, B. Plant guttation water as a potential route for pesticide exposure in honey bees: A review of recent literature. *Apidologie* **2018**, *49*, 637–646. [CrossRef]

57. Di Pasquale, G.; Alaux, C.; Conte, Y.L.; Odoux, J.F.; Pioz, M.; Vaissière, B.E.; Belzunces, L.P.; Decourtye, A. Variations in the availability of pollen resources affect honey bee health. *PLoS ONE* **2016**, *11*, e0162818. [CrossRef] [PubMed]

58. Pechhacker, H.; Praznik, W.; Klaus, J. Untersuchungen über das Zuckerspektrum in Honigblaseninhalt und Honig. *Apidologie* **1990**, *21*, 447–455. [CrossRef]

59. Imdorf, A.; Bogdanov, S.; Kilchenmann, V. Zementhonig im Honig- und Brutraum—Was dann?—Schweizerisches Zentrum für Bienenforschung. *Schweiz. Bienenztg.* **1985**, *108*, 534–544.

60. Rundlöf, M.; Andersson, G.K.S.; Bommarco, R.; Fries, I.; Hederström, V.; Herbertsson, L.; Jonsson, O.; Klatt, B.K.; Pedersen, T.R.; Yourstone, J.; et al. Seed coating with a neonicotinoid insecticide negatively affects wild bees. *Nature* **2015**, *521*, 77–80. [CrossRef]

61. Goulson, D.; Nicholls, E.; Botías, C.; Rotheray, E.L. Bee declines driven by combined Stress from parasites, pesticides, and lack of flowers. *Science* **2015**, *347*, 1255957. [CrossRef]

62. Rolke, D.; Fuchs, S.; Grünewald, B.; Gao, Z.; Blenau, W. Large-scale monitoring of effects of clothianidin-dressed oilseed rape seeds on pollinating insects in Northern Germany: Effects on honey bees (*Apis mellifera*). *Ecotoxicology* **2016**, *25*, 1648–1665. [CrossRef]

63. Requier, F.; Odoux, J.F.; Henry, M.; Bretagnolle, V. The carry-over effects of pollen shortage decrease the survival of honeybee colonies in farmlands. *J. Appl. Ecol.* **2017**, *54*, 1161–1170. [CrossRef]

64. Barroso-Arévalo, S.; Fernández-Carrión, E.; Goyache, J.; Molero, F.; Puerta, F.; Sánchez-Vizcaíno, J.M. High Load of Deformed Wing Virus and Varroa destructor Infestation Are Related to Weakness of Honey Bee Colonies in Southern Spain. *Front. Microbiol.* **2019**, *10*. [CrossRef]

65. Moosbeckhofer, R.; Köglberger, H.; Derakhshifar, I.; Morawetz, L.; Boigenzahn, C.; Oberrisser, W. *Varroa-Bekämpfung einfach-sicher-erfolgreich. 2. Völlig Neu Bearbeitete Auflage*; Biene Österreich: 2015. Available online: https://cdn.netletter.at/imkerbund/media/download/2016.02.09/1455008954025576.pdf?d=Varroabroschuere_Neu.pdf&dc=1455009208 (accessed on 7 February 2020)

66. Calderone, N.W. Evaluation of Drone Brood Removal for Management of *Varroa destructor* (Acari: Varroidae) in Colonies of *Apis mellifera* (Hymenoptera: Apidae) in the Northeastern United States. *J. Econ. Entomol.* **2005**, *98*, 645–650. [CrossRef]

67. Lodesani, M.; Franceschetti, S.; Dall'Ollio, R. Evaluation of early spring bio-technical management techniques to control varroosis in Apis mellifera. *Apidologie* **2019**, *50*, 131–140. [CrossRef]

68. Charrière, J.D.; Imdorf, A. Oxalic acid treatment by trickling against Varroa destructor: Recommendations for use in central Europe and under temperate climate conditions. *Bee World* **2002**, *83*, 51–60. [CrossRef]

69. Rademacher, E.; Harz, M. Oxalic acid for the control of varroosis in honey bee colonies—A review. *Apidologie* **2006**, *37*, 98–120. [CrossRef]

70. Colin, T.; Lim, M.Y.; Quarrell, S.R.; Allen, G.R.; Barron, A.B. Effects of thymol on European honey bee hygienic behaviour. *Apidologie* **2019**, *50*, 141–152. [CrossRef]

Article

Monitoring the Field-Realistic Exposure of Honeybee Colonies to Neonicotinoids by An Integrative Approach: A Case Study in Romania

Eliza Căuia *, Adrian Siceanu, Gabriela Oana Vişan, Dumitru Căuia, Teodora Colţa and Roxana Antoaneta Spulber

Institute for Beekeeping Research and Development, 013975 Bucharest, Romania;
adrian.siceanu@icdapicultura.ro (A.S.); gabrielaoana81@gmail.com (G.O.V.);
dumitru.cauia@icdapicultura.ro (D.C.); teodora.colta@icdapicultura.ro (T.C.);
roxana.spulber@icdapicultura.ro (R.A.S.)
* Correspondence: eliza.cauia@icdapicultura.ro

Received: 12 November 2019; Accepted: 30 December 2019; Published: 6 January 2020

Abstract: Honeybees (*Apis mellifera* L.) are excellent biosensors that can be managed to collect valuable information about environmental contamination. The main objective of the present study was to design and apply an integrative protocol to monitor honeybee colony activity and sample collection by using electronic technologies combined with classical methods in order to evaluate the exposure of honeybees to the neonicotinoids that are used in melliferous intensive crops. The monitored honeybee colonies were especially prepared and equipped to maximize their chances to collect representative samples in order to express, as well as possible, the pesticide residues that existed in the targeted crops. The samples of honey, pollen and honeybees were collected, preserved and prepared to fulfill the required quality and quantity criteria of the accredited laboratories. In total, a set of fifty samples was collected from fields, located in different areas of intensive agriculture in Romania, and was analyzed for five neonicotinoids. The obtained results show that 48% of the total analyzed samples (n = 50) contained one or more detected or quantified neonicotinoid residues. The main conclusion is that the proposed approach for sample collection and preparation could improve the evaluation methodologies for analyzing honeybees' exposure to pesticides.

Keywords: corn; honeybee colony; monitoring hive; neonicotinoids; oilseed rape; sunflower

---

1. Introduction

The insect pollination is one of the most important services that sustains biodiversity and food production. Out of the most important commercial crops 84% are insect pollinated [1]. Honeybees (*Apis mellifera* L.) are very important for increasing the quality of fruits and seeds of many wild and cultivated plant species [2,3]. For example, older research in Romania [4] showed the following pollinators' participation in stone tree pollination: honeybees—76.6%; bumblebees—7.6%; flies—3.7%; ants—3.6%; beetles—3.5%; wild bees—2.5%; wasps—0.5%; and other insects—2%.

Due to their complex biology (social life, reproduction, nutrition etc.) and intimate connection to climatic and vegetation conditions, honeybees are a natural biosensor of environmental quality [5–9]. Some of their special behaviors make honeybees a special pollinator: they exclusively feed on nectar and pollen; there is a high number of individuals in a colony, which leads to large quantities of food storage; they have the big ray of forage flight (0–5 km); their "flower fidelity" behavior makes them an efficient pollinator for a certain plant species at a certain moment; they visit many flowers based on the quality and quantity of nectar secretion; they have good orientation, memory and communication regarding food sources; they thermoregulate their nests, which helps overwintering; and they have the possibility to manage and transport their colonies to different crops.

In the context of developing technologies based on electronic sensors combined with the Internet of Things (IoT) and big data storage, honeybees could become even better suited to be equipped and used to monitor different physical and biological aspects of their colonies in order to help beekeepers' management, to avoid high honeybee losses, and to understand the different environmental factors that are causing the declining of their species [9–16]. Thus, from the perspective of managing qualities and their sensitivity to environmental contaminants, the honeybee colony represents an important bio tool to understand and even quantify the impact of different aggressive factors that intersect with its complex social life.

Honeybee depopulation and mortalities in the past two decades have become very important worldwide issues. Though the causes of these issues are quite diverse [17–19], there have been numerous articles on the environmental factors and pesticides. A series of research studies showed that in experimental and field conditions, honeybees' exposure to treated crops registered the following levels of residues: thiacloprid residues were in averages of 75.1 ng/g in pollen and 6.5 ng/g in honey, clothianidin residues were in averages of 9.4 ng/g in pollen and 1.9 ng/g in honey, and imidacloprid residues were in average of 19.7 ng/g in pollen and 6 ng/g in honey [20]. Thiacloprid was also frequently detected in honey samples up to concentrations of 200 ng/g [21], which is the maximum residue level that is accepted for human consumption. In another study [22] the occurrence of pesticides was more frequent in pollen and beeswax, and imidacloprid and fipronil were detected mostly in all matrices. A recent study that was focused on worldwide honey samples [23] showed that 75% of honey samples contained at least one neonicotinoid in quantifiable amounts, the total concentration of the measured neonicotinoids being 1.8 ng/g on average. In another piece of research [24], 97% of neonicotinoids found in pollen came from wildflowers, which grow near treated crops. The presence of more than one pesticide from different categories, in the same sample, is another issue that has been highlighted in much research worldwide that has shown that honeybees are chronically exposed to neonicotinoids and fipronil, and their metabolites in the 1–100 ng/g (ppb) range [20,25,26]. In one study [27], sunflowers and corn tassels contained values of 10 ng/g of imidacloprid in average, which explained why the pollens from these crops were contaminated at levels of a few ng/g. Another study [28] showed that the concentrations of imidacloprid found in sunflower and corn pollens collected in the pollen traps were, respectively, about 1.5 and 4.5 times less than the concentration of imidacloprid found in the same types of pollen that were directly collected from flowers, thus highlighting the potential hazard of neonicotinoids to honeybees through contaminated pollen and nectar. One survey [29] showed that colonies located in a corn-dominated area registered greater colony mortality by 3.51 times more than in corn crop-free locations. In the same study, it was shown that 54% of analyzed samples contained clothianidin, and 31% contained both clothianidin and thiamethoxam.

Studies have shown that these substances affect the honeybee organism at nanograms levels (1 ng = 10^{-9}), with 0.10 ng imidacloprid/bee being the lowest concentration that had detrimental effects on honeybees in laboratory conditions [30].

Following the scientific data of numerous studies regarding their toxicity by lethal and sublethal effects, based on European Food Safety Authority scientific reports [31], the European Union (EU) (2018/783/784/785/29.05.2018) banned the use of three neonicotinoids (imidacloprid, clothianidin, and thiamethoxam) in fields. However, a series of countries were approved to use these substances by emergency authorization, Romania being the country which continuously used the forbidden neonicotinoids, at the country level, based on emergency authorizations.

In this context, it is important to mention that Romanian agriculture is one of the main economic sectors that is continuously developing thanks to the favorable geographic and climatic conditions. This situation has led to an increase of land surfaces that are cultivated with industrial crops (sunflower, rape, and corn) and their inputs. In the same time, beekeeping has been a very important sub-domain of agriculture that is favored by natural conditions and stimulated by the organizational national structure under Romanian Beekeepers Association, founded in 1958.

Since Romania's admission into the EU (2007), the developing beekeeping sector has been encouraged by the national beekeeping program and some other agricultural programs that are ruled by European and national legislation. As a result, the number of hives has constantly increased, with the total number of hives being 11.7% of the total EU number with a production of honey of 13% of the total EU honey production in 2018 (these data were published at https://ec.europa.eu/agriculture/honey/programmes_en).

If rape and sunflower are very well known in importance for beekeeping, corn crops (*Zea mays*) are not so well highlighted as they are considered a wind pollinated crop. However, even pollen morphology reflects an adaptation to wind pollination, as its nutritional properties make it an important attractant for honeybees and other pollinating insects. Its male flower (the tassel) offers large amount of pollen that can be very attractive during the period of sunflower honey flow when crops are nearby. This source of pollen is of great importance as its flowering period overlaps on that when winter honeybees' generations are reared, so it substantially contributes to the quality of wintering, both by honeybee quality as well as by pollen storage that is consumed in the early stage of brood rearing in the next season. As a consequence, the contamination of corn pollen with neonicotinoids may have a great negative impact on honeybees [32–37] in the sunflower honey flow period, as a lot of corn fields are closer to the sunflower crop fields—this is a reason why corn crops were included in this study.

Taking into account policy context and certain beekeepers' complaints regarding honeybees' depopulations [38], a research project was funded in October 2017–October 2018 by the Ministry of Agriculture and Rural Development for the first time in Romania. The aim of this project was to establish the realistic-field exposure levels to neonicotinoids in certain areas that are intensively cultivated with oilseed rape, corn and sunflower. To carry out this research, an integrative approach was developed in order to follow up the honeybee colonies and to collect and prepare representative samples in order to evaluate the exposure of honeybees to the neonicotinoids used in intensive crops.

2. Materials and Methods

This research was carried out in the following phases:

(1) Field identification in different agricultural areas.
(2) Honeybee colony preparation and transportation to different envisaged crop fields.
(3) Sample collection and specific preparation.
(4) Sample preservation, codification, packing, and sending to the accredited laboratories according to specific requirements.
(5) The neonicotinoid analyses and results reports performed by the accredited laboratories.

2.1. Field Identification in Different Agricultural Areas

To implement the protocol and carry out the analyses regarding the exposure of honeybee colonies to neonicotinoids, three fields were selected in different areas of intensive agriculture in the southeastern and eastern parts of Romania, being provided by two different agricultural research stations and one institute that belongs to the Romanian Academy for Agricultural and Forestry Sciences "Gheorghe Ionescu Sisesti". These were located in Neamt county (Statiunea de Cercetare Dezvoltare Agricola—SCDA Secuieni, 46°51′45″ N, 26°49′42″ E), Arges county (Statiunea de Cercetare Dezvoltare Agricola—SCDA Albota, 44°46′54″ N, 24°49′31″ E), and Calarasi county (Institutul National de Cercetare Dezvoltare Agricola—INCDA Fundulea, 44°27′10″ N, 26°30′55″ E). Besides these, one farmers' association was involved in this monitoring study: The Corn Producers' Association (Asociatia Producatorilor de Porumb din Romania—APPR) located in Ialomita county Mihail Kogalniceanu, Tandarei (44°40′39.67″ N, 27°42′14.71″ E) (Figure 1).

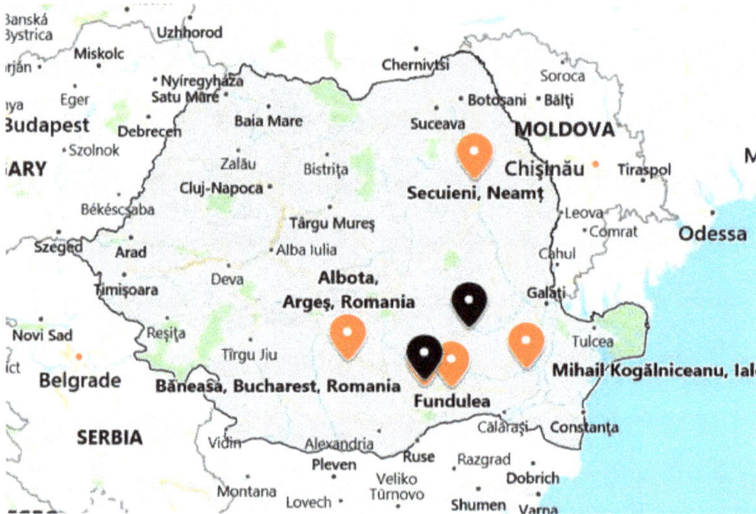

Figure 1. Monitored locations by honeybee colonies to field-realistic exposure to neonicotinoids in Romania.

These centers provided different surfaces of land (a minimum of 1 ha and a maximum of 100 ha) that were cultivated with targeted crops (rape, corn and sunflower) and treated with field-prescribed doses of active substances/products by seed dressing, regarding the three neonicotinoids (imidacloprid, clothianidin, and thiamethoxam) that are the subject of interdiction in the European Union. In addition to these locations, a series of samples were collected from two apiaries belonging to the Institute for Beekeeping Research and Development Bucharest, apiaries being located in different areas in southeastern Romania (Baneasa-Bucuresti—44°29′33″ N, 26°04′45″ E, Buzau—45°10′ N, 26°49′ E), as well as from private beekeepers (Fundulea—44°27′10″ N, 26°30′55″ E, Otopeni—44°32′ N, 26°6′ E,) (Figure 1), in order to evaluate the presence of neonicotinoids in different areas with intensive agriculture but without any information about the use of phytosanitary treatments.

The experimental honeybee colonies were located near the treated crops (0–100 m), depending on the configuration of the land, at approximatively 10% blooming, in order to attract the honeybees for nectar or pollen collection and initiate the "fidelity flower" behavior.

2.2. Honeybee Colony Preparation and Transportation to Different Envisaged Crop Fields

In order to monitor the honey/pollen flows and collect representative samples, the experimental honeybee colonies were prepared. In this regard, every location was supplied with two honeybee colonies (normal colonies with queen, brood and food storages), established on 10 Dadant frames each, well covered with honeybees, and equipped with entrance type pollen collectors and foundation frames for honey collection in order to let honeybees build combs in the honey flow conditions to decrease the risk of contaminating the collected honey by older combs.

Out of these colonies, one honeybee colony per each location was equipped with an electronic hive (Simbee®, http://www.simbee.ro/) that consisted of a special module of sensors and data collectors: two ambient temperature and humidity sensors, an internal sensor for honeybee colony temperature, and a weight sensor scale.

The monitoring hives transmitted the collected data to its database every 10 min, thus registering the activity of bee colonies regarding the monitoring parameters and helping to understand the existence or lack of honey flows in the flight area, as well as suspicions about a possible decline of population and its development status that are connected with phytosanitary treatments.

2.3. Sample Collection and Specific Preparation

2.3.1. The Honeybee Samples

These samples were collected from dying or live honeybees, depending on the encountered situation. The dying honeybees were collected out from the front of hive and, when faced a lack of dead or dying honeybees, we collected live honeybees (foragers) from the entrance, following the honey or pollen flow, by using a car aspirator. The samples were immediately confined to a bag and put in a car freezer.

As during the experiments, acute and lethal effects on honeybees were not directly observed in most of the cases, the honeybee sampling consisted of the collection of live forage honeybees from the entrance of the hives in order to increase the probability of finding neonicotinoids in honeybees that carry freshly contaminated nectar or pollen.

2.3.2. The Honey and Pollen Samples

After 7–10 days from the beginning of the honey flow, honey samples were collected from specially prepared and introduced frames in the monitored hives that were preserved in refrigerators.

The collected honeybee pollens were taken out from collectors and preserved in specific low temperature conditions (between −10 and 4 °C) daily, depending on local situation, until their mono-floral analyses and special preparing samples, which was the case for samples sent to the French Agency for Food, Environmental and Occupational Health and Safety (ANSES) laboratory.

In order to identify the source of different contaminants by using the honeybee as a sampler, one problem was represented by the big ray of forage flight from the hive (0–5 km), which covers a big surface of land and seldom obtains multi-floral products, e.g., honey, pollen, and beebread. In the correct monitoring of a specific crop, the problem lies in collecting its specific mono-floral samples, as this has a big impact on residue data identification and interpretation.

Thus, to understand the floral componence and to establish the mono-floral honeys from the targeted field, the samples collected out from the rape and sunflower crops were analyzed in the framework of the chemistry laboratory of the Institute for Beekeeping Research and Development, based on specific standardized methods of melissopalinology used in the evaluation of honey types (Table 1). These analyses are very important in the context of the neonicotinoid residues analysis because they confirm whether the samples are sufficiently relevant for the purpose of the study.

Due a lack of mono-floral envisaged samples because of climatic conditions (drought or heavy rains specific to the 2018 season) in the present study, we also used honey collected during the studied crop flowering that was classified as multi-floral honeys but also contained the targeted honey in different percentages.

Regarding the mono-floral pollen samples and taking into account the variability of pollens usually collected by honeybees, most of the pollens collected by specific entrance collectors in the rape and sunflower period were multi-floral. To have relevant, envisaged mono-floral pollen samples for neonicotinoid analyses, based on the minimum required quantities (when possible), we manually selected the pollen pellets of rape, corn and sunflower (Figure 2a,b) in the laboratory conditions based on the pellet aspect and color, with the selection being randomly confirmed by microscopy based on pollen characteristics (Figure 3a–c).

Pollen mono-floral selection, a time consuming activity, was possible only when the minimum sample quantity requested by laboratory was low. The minimum quantities of the samples requested by the two laboratories were a minimum of 10 g at the ANSES laboratory and a minimum of 250 g at Quality Services International (QSI) laboratory. The minimum quantity of 10 grams (e.g., ANSES laboratory) permitted a better approach in the sample preparation, thus providing a better analysis of the pollen origin regarding the plant species. The process of collection and selection of representative samples was a key stage in the neonicotinoid identification and quantification.

Table 1. The obtained results on different types of honey samples, following the mellissopalinological analyses.

Honey Sample (According to Samples Codification)	Honey Collected Source	Pollen Grains from the Envisaged Crop	Other Significant Pollens Found (Family, Genus)	Result
R-M-F-1	Oilseed rape	81% *Brassica napus*	*Prunus, Malus, Gleditsia, Salix*	Rape honey (mono-floral)
R-M-FB-1	Oilseed rape	88% *Brassica napus*	*Prunus, Malus, Apiaceae*	Rape honey (mono-floral)
R-M-S-1	Oilseed rape	45% *Brassica napus*	*Prunus, Malus, Apiaceae, Taraxacum*	Rape honey (mono-floral)
R-M-O-1/2	Oilseed rape	90% *Brassica napus*	*Prunus, Malus, Taraxacum*	Rape honey (mono-floral)
FS-M-A-1/2	Sunflower	10% *Helianthus annuus*	*Castanea, Apiaceae, Onobrychis, Trifolium, honeydew* elements.	Multi-floral honey
FS-M-S-1/2	Sunflower	45% *Helianthus annuus*	*Cirsium, Gramineae, Tilia, Castaneae*	Sunflower honey (mono-floral)
FS-M-T-1/2	Sunflower	37% *Helianthus annuus*	*Ilex, Gramineae, Cirsium, Apiaceae*	Multi-floral honey

(a)   (b)

Figure 2. Example of honeybee multi-floral pollen samples collected by honeybees in the two studied periods. The rape, sunflower and corn pollens pellets were selected in the laboratory when preparing the samples to be analyzed. (**a**) A pollen sample collected during the oilseed rape (*Brassica napus*) blooming. One can notice an amount of approximately 50% oilseed rape pollen, (indicated by red circles). (**b**) A pollen sample collected during the sunflower (*Helianthus annuus*) and corn (*Zea mays*) blooming. One can notice an amount of approximately 60% sunflower pollen (orange) versus 40% corn pollen (yellow). Photos© Institute for Beekeeping Research and Development, Bucharest.

Following these selection steps, 50 samples were prepared and sent to the laboratories, as follows: honeybees—10 samples; honey—15 samples; and pollen—25 samples.

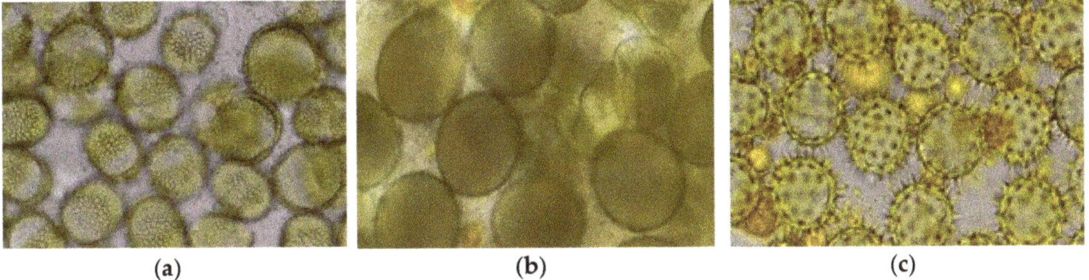

(a)   (b)   (c)

Figure 3. Microscopic view of the pollen grains used to confirm the studied selected samples. (**a**) The oilseed rape (*Brassica napus*) pollen had a ~24 μm grain with a round tricolpate form and grooves at the rounded tip, covered with a membrane, sometimes with small granular residues; the exine had a finely cross-linked surface. (**b**) The corn (*Zea mays*) pollen grain was a ~99 μm granule with a single porous opening, the exine and the intine being thin membranes and the cytoplasm appearing granular with numerous small starch formations. (**c**) The sunflower (*Helianthus annuus*) pollen grain was ~35 μm and was a trickled grain, having a surface covered in thin and long thorns that protruded to the pores. The used equipment: Olympus, Quick Photo camera 3.1., ×200: Photos© Institute for Beekeeping Research and Development, Bucharest.

2.4. Sample Preservation, Codification, Packing and Sending to the Accredited Laboratories According to Specific Requirements

After the collection and transportation to the central laboratory of the Institute for Beekeeping Research and Development, all the samples were preserved at −18 °C.

For identification, samples were coded with a unique code, e.g., R-P-F-2 = Rape-Pollen-Fundulea-2nd sample (R = rape; FS = sunflower; P = corn; M= honey; P= pollen; and A= honeybees). The samples, which were prepared according to the specific requirements of analyzing laboratories, were packed in special containers and sent to the following two European accredited laboratories: thirty samples were sent to an EU Reference laboratory—French Agency for Food, Environmental and Occupational Health and Safety (ANSES), France, and 20 samples were sent to Quality Services International (QSI), Germany. The shipping was performed in special freezing packing for biological material.

The two laboratories were chosen in order to diversify and better understand the requirements of the accredited laboratories regarding the preparation of samples for analyses. Generally, the samples were distributed between the two laboratories depending on the collected quantity correlated with melissopalinological analysis.

The results of neonicotinoid analyses performed by the accredited laboratories are presented in Tables 2 and 3, together with their level of detection and quantification on five neonicotinoids—acetamiprid, clothianidin, thiamethoxam, imidacloprid and thiacloprid. The analyses were done with the liquid chromatography method coupled with tandem mass spectrometry (LC-MS/MS) at both laboratories.

Table 2. The results on the neonicotinoid residues found in samples collected from different crops and locations in Romania (2018).

			Neonicotinoid Residues in Samples Collected from Different Crops								
			Oilseed Rape							Corn	Sunflower
No.	The Active Substance	Locations	Honeybees (ng/bee [1])		Honey (ng/g [2])			Pollen (ng/g)			Pollen (ng/g)	Honey (ng/g)
			ANSES		ANSES		QSI	ANSES		QSI	ANSES	ANSES
			<LOQ	>LOQ	<LOQ	>LOQ	>LOQ	<LOQ	>LOQ	>LOQ	>LOQ	<LOQ
1	Imidacloprid	Albota	-	-	-	-	-	-	11.7	-	-	-
		Secuieni	-	-	-	-	-	-	-	-	1.1	-
		Tandarei	<0.05	-	<1.0	-	-	-	10.6	-	1.4	<1.0
		Fundulea	-	0.1	-	-	-	-	31.1	13.0	1.2	-
		Baneasa	-	0.1	-	-	-	-	-	-	-	-
2	Thiamethoxam	Albota	-	-	-	-	-	<0.5	-	-	-	-
		Tandarei	-	-	-	-	-	<0.5	-	-	-	-
		Fundulea	-	-	-	-	-	<0.5	-	-	-	-
3	Acetamiprid	Albota	-	-	-	1.8	-	-	-	-	-	-
		Secuieni	-	-	-	-	-	-	-	-	-	<1.0
		Tandarei	-	-	<1.0	-	-	-	9.9	-	-	-
		Fundulea	-	-	-	-	-	<1	-	-	-	-
		Otopeni PB	-	-	-	7.3	-	-	-	-	-	-
4	Thiacloprid	Albota	-	-	-	3.6	-	-	795.1	11.0; 13.0	-	<1.0
		Secuieni	-	-	-	2.3	-	-	-	-	2.0	-
		Tandarei	-	-	-	3.7	1.2	-	131.8	11.0	-	-
		Fundulea	-	-	-	-	-	-	2.4	-	-	-
		Buzau	<0.05	-	-	-	-	-	-	-	-	-
		Otopeni PB	-	-	<1	-	1.4	-	-	-	-	-
		Fundulea PB	-	-	-	-	3.1	-	-	-	-	-

[1] ng/bee = nanogram per bee; [2] ng/g = nanogram per gram (= ppb, parts per billion); a dash means either not sampled or not found.

Table 3. The limit of detection and quantification of the involved laboratories, as well as the maximum residue limit in conformity with European legislation.

The Analyzed Substance	Honeybees Samples		Honey Samples				Pollen Samples			
	LOD [3] (ng/bee)	LOQ [4] (ng/bee)	LOD (ng/g)	LOQ (ng/g)		MRL [5] (ng/g)	LOD (ng/g)	LOQ (ng/g)		MRL (ng/g)
	ANSES	ANSES	ANSES	ANSES	QSI		ANSES	ANSES	QSI	
Acetamiprid	0.03	0.10	0.3	1.0	1.0	50.0	0.3	1.0	10.0	50.0
Clothianidin	0.015	0.05	1.5	4.0	5.0	50.0	0.15	0.5	10.0	50.0
Thiamethoxam	0.015	0.05	1.5	4.0	1.0	50.0	0.15	0.5	10.0	50.0
Imidacloprid	0.015	0.05	0.3	1.0	5.0	50.0	0.15	0.5	10.0	50.0
Thiacloprid	0.015	0.05	0.3	1.0	1.0	200.0	0.15	0.5	10.0	200.0

[3] LOD = limit of detection; [4] LOQ = limit of quantification; [5] MRL= the maximum residue level permitted for human consumption according to European legislation.

3. Results

The 2018 beekeeping season was generally a weak active season for Romania in terms of honey production, (via oilseed rape and sunflower), this situation being registered in the monitored locations by means of the electronic hives. The weak beekeeping season had a negative impact on honey sample collection. The weight gain represented the first indicator of honeybee colony activity and, mainly, of the existence of a nectar flow from targeted fields that is necessary for honey sample collections. This evaluation offered preliminary information about the probability of collecting samples from the targeted fields, but this information needed to be correlated and confirmed by melissopalinological studies on collected samples. The weight gain in the monitored hives is shown in the following images (Figures 4 and 5):

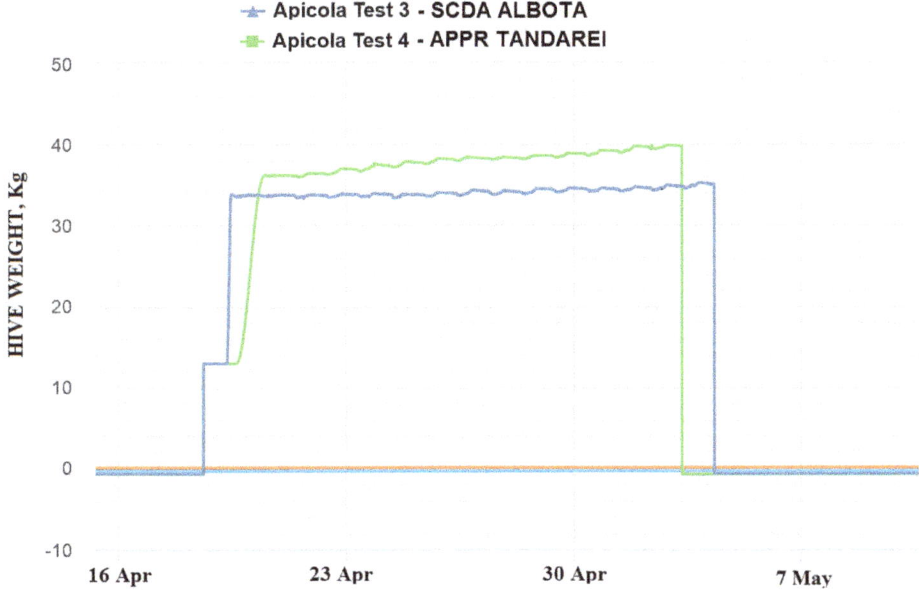

Figure 4. The weight gain (kg) monitored by experimental electronic hives in the rape honey flow in two locations (Albota and Tandarei) from 20 April to 5 May 2018.

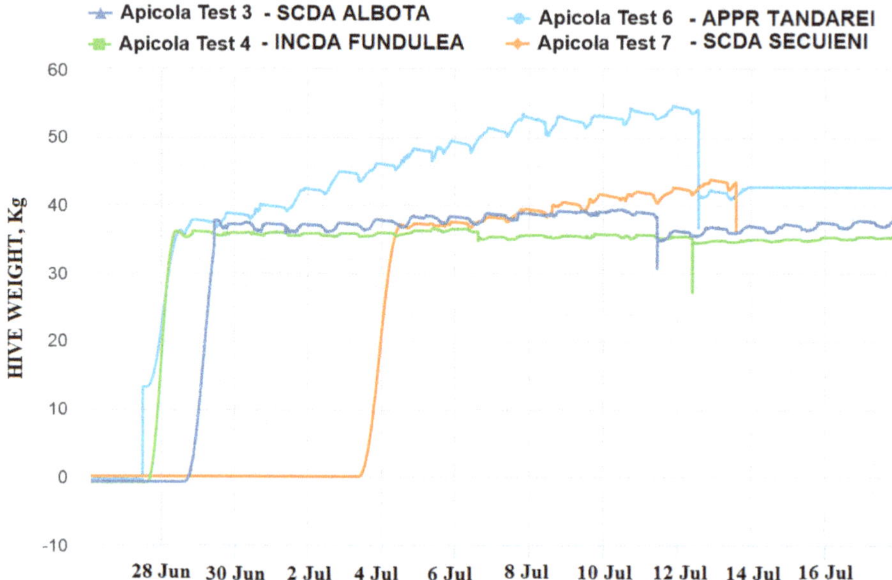

Figure 5. The weight gain (kg) monitored by experimental electronic hives in the sunflower honey flow in four locations from 29 June to 14 July 2018.

As can be noticed from the gain weight registered by the electronic hive, the general activity of honeybee colonies and the development of the colonies in honey flow conditions were relatively low, with the causes being not well understood.

Concerning the results on honeybees, it is important to mention that in general, in the Fundulea, Tandarei and Baneasa locations, a weak activity was noticed. Additionally noticed were signs of depopulations and/or honeybees in front of the hives with specific symptoms of acute toxicity (walking on the ground, paralysis, and dying) during the oilseed rape honey flow period. The results showed that imidacloprid was present in a concentration of 0.1 ng/bee, as well as under the limit of quantification (LOQ) in these locations.

The presented protocol that was used to collect and prepare honey and pollen samples for neonicotinoid residues analyses, included an important step: melissopalinological analyses for the identification the mono-floral honeys (Table 1) and mono-floral pollens selection. Following these preliminary analyses, all rape honey samples were framed in the standards of mono-floral honeys, and the results showed that these samples were set according to the standard internal laboratory references for rape honey (minimum 40% rape pollen grains). However, in the case of sunflower honeys, only one of the collected samples was framed in the internal standards of typical sunflower honey.

These results indicate that the analyzed honey samples in the case of multi-floral honeys (e.g., sunflower honey) could not correctly reflect the residues of neonicotinoids found in the envisaged flora, and this led to their sub-evaluation.

The results on the neonicotinoid residues analyzed on different matrices and laboratories are highlighted in Table 2.

The raw data show that 48% of the total samples ($n = 50$) sent to the two laboratories contained one or more, detected or quantified, neonicotinoid residues, with the quantifiable residues being found in 38% of samples.

It can be emphasized that, in the case of the EU reference laboratory (ANSES), 43.3% of the total analyzed samples ($n = 30$) contained registered, quantifiable residues of one or more neonicotinoids, and 33.3% contained registered, detectable amounts of one or more residues of neonicotinoids. In the QSI laboratory, 30% of the total analyzed samples ($n = 20$) contained registered, quantifiable residues.

One very important aspect needs to be mentioned regarding the analytical references of the two laboratories (Table 3) for the residue quantification limit (LOQ), which could influence the results and their interpratation; for example, in the honey analyses, the different residue LOQs (ANSES = 1.0–4.0 ng/g, QSI = 1.0–5.0 ng/g) need to be analyzed on different active substances between the two laboratories, while for the pollen analyses, the LOQs (ANSES = 0.5–1 ng/g, QSI = 10.0 ng/g) are more favorable in ANSES laboratory.

Taking into account the LOQ of the residue analysis in the involved laboratories, the following neonicotinoids were found in quantifiable levels: imidacloprid in honeybee samples, acetamiprid and thiacloprid in honey samples, and acetamiprid, imidacloprid and thiacloprid in pollen samples.

The minimum–maximum levels of the quantified neonicotinoids were:

(1)   0.1 ng/bee imidacloprid in honeybee samples in the oilseed rape blooming period.
(2)   1.2–3.7 ng/g thiacloprid in oilseed rape honey samples.
(3)   1.8–7.3 ng/g acetamiprid in oilseed rape honey samples.
(4)   1.3–31.1 ng/g imidacloprid in oilseed rape pollen samples.
(5)   1.1–1.4 ng/g imidacloprid in corn pollen samples.
(6)   2.0 ng/g thiacloprid in corn pollen samples.
(7)   1.1–795.1 ng/g thiacloprid in oilseed rape pollen samples.

Regarding the three neonicotinoids (imidacloprid, clothianidin and thiamethoxam) that are banned at European level but are the subject of derogation in Romania, the following percentages of residues were found on different matrices and crops in the two laboratories:

(1)   In honeybees collected from oilseed rape crops, imidacloprid was in 50% of the total collected samples ($n = 6$) and was found in 33.3% of the samples, being quantified at 0.1 ng in the ANSES laboratory;
(2)   In the honey collected from oilseed rape crops, imidacloprid was found under the LOQ in 16.6% of the total samples ($n = 6$) analyzed in the ANSES laboratory;
(3)   In honey collected from sunflower crops, imidacloprid was found under the LOQ in 25.0% of the total samples ($n = 4$) analyzed in the ANSES laboratory;
(4)   In pollen collected from oilseed rape crops, imidacloprid was found in 16.6% ($n = 1$, 13 ng/g) of the total samples ($n = 6$) analyzed in the QSI laboratory;
(5)   In pollen collected from oilseed rape crops, imidacloprid was found in concentrations of between 10.6 and 31.1 ng/g in 100% of the total samples ($n = 3$) analyzed in the ANSES laboratory;
(6)   In pollen collected from corn crops, imidacloprid was found in concentrations between 1.1 and 1.4 ng/g in 100% of the total samples ($n = 3$) analyzed in the ANSES laboratory.

Out of all analyzed active substances, clothianidin was not found in any sample.

Concerning the monitored crops, the percentages of samples with one or more quantifiable neonicotinoid residues in different matrices are presented as it follows:

(1)   In the oilseed rape crops, neonicotinoid residues were found in 33.3% of the honeybee samples ($n = 6$), in 87.5% of the honey samples ($n = 8$), and in 77.7% of the pollen samples ($n = 9$).
(2)   In the corn crops, neonicotinoid residues were found in 100% of the pollen samples ($n = 3$).

With only one exception found in oilseed rape pollen (795.1 ng/g thiacloprid), all residues were under the maximum residue limit (MRL) established for human consumption.

Looking to the percentages of samples with one or more neonicotinoid residues over the detection limit in different matrices and crops, the situation is presented as follows:

(1) In the oilseed rape crops, one or more neonicotinoid residues over the detection limit were found in 66.6% of the honeybee samples ($n = 6$), in 87.5% of the honey samples ($n = 8$), and in 77.7% of the pollen samples ($n = 9$).

(2) In the corn crops, one or more neonicotinoid residues over the detection limit were found in 100% of the pollen samples ($n = 3$).

(3) In the sunflower crops, one or more neonicotinoid residues over the detection limit were found in 42.8% of the total honey samples ($n = 7$).

One can notice the presence of neonicotinoids in detectable amounts in three of the four sunflower honey samples analyzed in the ANSES laboratory, while in sunflower pollen from the same location (Albota, Tandarei, and Secuieni), neonicotinoids were not detected, even in the pollen samples that were mono-floral selected. An inverse situation was noticed in one oilseed rape sample analyzed in the same laboratory (ANSES), where neonicotinoid residues were not identified in honey (R-M-F-1) from the same location (Fundulea), but they were very well identified in a pollen sample (R-P-F-1). This result shows the importance of analyzing residues in both matrices of honey and pollen collected from the same hive.

Another aspect of residue analysis is related to the presence in detectable and/or quantifiable amounts of different active substances of neonicotinoids in the same sample that could conduct the so-called "cocktail" effect. In this study, this was the case of oilseed rape honey (50%) and pollen samples (100%), as well as in corn samples (33.3%), though only in the samples analyzed in the ANSES laboratory.

Another important aspect that is worth mentioning is that in every sample preparation, the corn and sunflower pollens pellets were separated from the same multi-floral sample with different levels of mixtures with other pollens (not analyzed). Out of the obtained results, one can highlight that the neonicotinoid residues found in corn pollens did not influence the sunflower pollen residues because all corn pollens registered different levels of residues and sunflower pollen was free of residue from an analytical point of view. This fact indicates us that neonicotinoids from a pollen pellet do not contaminate other pollen pellets.

Taking into account the eight locations where the samples were collected from, the results showed that every location had at least one sample with detected or quantified neonicotinoids in honey or pollen, so the distribution of neonicotinoids in the environment was relatively large in the areas with intensive agriculture.

4. Discussion

The present study aimed to improve the evaluation methodologies for analyzing honeybees' exposure to pesticides in order to better identify and quantify the contamination of honeybees' food resources from the main treated crops, as well as to have a first image of their presence and quantification in Romania in some treated crops. Thus, the most important levels of neonicotinoid residues, in the vegetation conditions of the 2018 season, were identified in oilseed rape honeybees, honey, and pollens, as well as in corn pollen.

In view of these results, the analysis of neonicotinoids in these matrices is very important for establishing a basic exposure level of honeybees to pesticides in a specific area at a certain moment.

The identification and quantification of neonicotinoids in any sample could be a combined result of many factors, but the collection and sample preparation until their analysis, as well as the analysis methodology with the lowest LOD (limit of detection) and the lowest LOQ, are of great importance.

In order to identify and quantify the neonicotinoids, the obtained results in the two laboratories also showed the importance of the mono-floral sample preparation of honeys and pollens.

One interesting finding concerned the presence of different neonicotinoids in detectable and/or quantifiable amounts in all matrices that were obtained from oilseed rape crops, while some neonicotinoids (acetamiprid, imidacloprid and thiacloprid) were only found in detectable amounts in sunflower honey. The low level of neonicotinoid residues obtained in some sunflower honey samples can be explained by the fact that sunflower honey is actually a multi-floral honey, so the contamination residues were diluted by different sources of nectars. However, the lack of neonicotinoids in the selected sunflower pollen samples remains questionable, and further research is necessary.

Following the identification of neonicotinoids (imidacloprid) in all the three corn pollen samples and in view of the fact that honeybees collect high quantities of corn pollen in areas with intensive crops, this study can be seen as offering an important overview of the importance of this crop for honeybee nutrition.

What is remarkable here, from the field observation, is the fact that corn tassels were intensely visited by honeybees, even when the nearby sunflower crops are in full bloom. This shows that corn tassels are an important source of pollen for honeybee colonies, as is sunflower pollen. For this reason, the corn crops represent important sources of nutrients, not only by guttation water but also by pollen, so its contamination with pesticides could affect honeybees during the whole vegetation period.

The obtained results are similar with those found in the literature [20–29,39] that have shown that the residues of neonicotinoids in honeys and pollens are found in the range of a few nanograms per gram. Their lethal or sublethal effects on honeybees depend on many factors [28] such as daily consumption, seasonal conditions, the activity and strength of the colony, the age and the duration of exposure, and health state. For example, the literature data show that the lethal toxicity of imidacloprid is in the range of a few picograms if ingestion is repetitive for minimum eight days and of a few nanograms if the ingestion is for one-to-two days [40]; however, its sublethal effects, such as learning and orientation ability modifications, could appear at a concentration of 0.1 ng/bee [30], which was the case in our study in real conditions.

Taking into account the toxic and cumulative effects of neonicotinoids, based on irreversible bind on the nicotinic acetylcholine receptors (nAChRs), as well as the toxic effects of their metabolites proven by research [41–45], the found concentrations pose serious risks to honeybee health in the short and long term. One such study [44] that was done over a short period of time (10–30 days) showed, by extrapolation, that the daily ingestion of about 0.005 ng/day of imidacloprid produced important lethal toxic effects (LT50) in 150 days. If a bee consumes around 0.02 g honey per day [28], a concentration of 0.25 ppb in honey (which is not a quantifiable amount (see LOQ, Table 3)) can cause long-term mortalities (over 150 days), as it happens in the winter. These low residue quantities can be consistently supplied by residues in storage pollens (beebread) in the late winter-to-early spring period when colonies begin rearing their brood, and this situation can explain the collapse of colonies due to a long exposure to sublethal effects.

Some research has gone even deeper, demonstrating that the use of neonicotinoids can lead to a wider range of sub-lethal effects on honeybees as a result of very low concentrations of neonicotinoids. Thus, if lethal acute effects can be rapidly noticed by the rapid decrease of the population or mortalities in front of the hive, sub-lethal effects can be difficult to observe, as colonies generally have problems of development or slow mortality for longer periods (autumn, winter, and early spring depopulation), inducing a variety of behavioral dysfunctions. Many of these dysfunctions affect orientation, memory, communication [46–48], foraging and flight [49,50], the olfactory sense [51], the glandular system and respiratory rhythm [52,53], reproduction [54,55], global temperature and metabolism [56,57], sensitivity to diseases, and immunity reduction [58–64].

Taking into account the high level of residue of thiacloprid found in one sample (four times more than the maximum residual limit for human consumption), it is important to show the risks that the EFSA mentioned in its report published in 2019 [65], such as: "delayed effects or relevant sub-lethal effects on bees at relatively low concentrations cannot be excluded" and "thiacloprid presents important risks for human health."

Honeybees are a very important biosensor that can be managed to obtain information about the environment. Nonetheless, the samples collected by honeybees need to be melissopalinologically analyzed when it comes to the necessity of analyzing a specific crop or plant species.

The collection, preparation and preservation of samples should be done so that to reflect the pesticide residues in nectar and pollens at the time of their collection by honeybees. This is of great importance in pesticide exposure studies in order to identify the real residues of neonicotinoids at detection or quantification levels and their risks to honeybees.

The monitoring of honey flows by electronic hives, even if not very important for sample collection and preparation, is the first indicator about honey flow, weather conditions, depopulations or other activities of honeybee colonies that give preliminary information on weight gain and honeybee populations (e.g., swarming). This basic electronic system could be completed by a specific electronic device that could measure, with high accuracy, entrance activity in order to highlight any slight modification in the number of foragers and their loss in the field over the normal levels of depopulation. In this sense, a series of new research projects are necessary to quantify the number of outgoing and incoming bees at the entrance and by these data processing measures in order to offer important information about abnormal depopulations and to send specific alerts in a useful time. Through this approach, it is possible to collect important information about realistic field depopulations and to collect relevant samples for neonicotinoid residues at depopulation moments in order to better analyze the impact of pesticide field exposure on honeybee health.

Thus, the registration of honey flow monitoring data by electronic means, the detection of general colony dynamic activities by electronic sensors, the collection and preparation of mono-floral honeys or pollens, the use of small samples for analyses in order to facilitate their mono-floral preparation for the further specialized laboratory analyses, and the good preservation of samples from the collection moment to the laboratory analyses all contribute to an effective system for neonicotinoid identification.

5. Conclusions

Considering the fact that, generally, neonicotinoid residues are found in honeybee colonies at low levels (e.g., ng/bee and ng/g), it is very important to have a reliable methodology to collect representative samples for neonicotinoids analyses.

In this regard, the good preparation and follow up of honeybee colonies and honey/pollen flows by using classic methods or techniques, new technologies (electronic sensors and the IoT), the validation of the mono-floral envisaged samples by specific mellissopalinological analyses, and their preservation at low temperatures from collection until neonicotinoid residues analysis are very important steps.

The presented approach for monitoring neonicotinoid residues in honeybee colonies could help to maximize the chances for their identification and quantification in different monitoring studies or evaluations, in order to better evaluate the exposure of honeybees to the neonicotinoids used in different real field melliferous crops.

As could be seen from the levels of neonicotinoid residues found in different locations in Romania, their presence in all the sampled locations correlated with the worldwide scientific evidence on their lethal or sublethal toxic effects in chronic exposure, demonstrate the existence of important risks for honeybee health and local beekeeping.

Author Contributions: Conceptualization, E.C., A.S., G.O.V. and D.C.; data curation, E.C. and A.S.; formal analysis, E.C., A.S., G.O.V., D.C., T.C. and R.A.S.; funding acquisition, E.C. and A.S.; investigation, E.C., A.S., G.O.V., D.C., T.C. and R.A.S.; methodology, E.C., A.S., G.O.V., D.C., T.C. and R.A.S.; project administration, A.S.; resources, E.C., A.S., G.O.V., D.C., T.C. and R.A.S.; validation, E.C., A.S., G.O.V., D.C., T.C. and R.A.S.; visualization, E.C., A.S., G.O.V. and D.C.; writing—original draft, E.C., A.S. and G.O.V. All authors have read and agreed to the published version of the manuscript.

Funding: This research was funded by Ministry of Agriculture and Rural Development, Romania, grant number ADER 4.1.5/2017.

Acknowledgments: We acknowledge administrative and technical support received by the Institute for Beekeeping Research and Development, Bucharest, Romania. The authors thank to the reviewers for the constructive analysis of the article.

Conflicts of Interest: The authors declare no conflict of interest. The funders had no role in the design of the study, in the collection, analyses, or interpretation of data, in the writing of the manuscript, or in the decision to publish the results.

Ethics Statement: We declare this study does not need approval of any institution and/or governmental agency that regulates research with animals.

References

1. Allsopp, M.H.; De Lange, W.J.; Veldtman, R. Valuing insect pollination services with cost of replacement. *PLoS ONE* **2008**, *3*, e3128. [CrossRef] [PubMed]
2. Corbet, S.; Williams, I.; Osborne, J. Bees and the pollination of crops and wild flowers in the european community. *Bee World* **1991**, *72*, 47–59. [CrossRef]
3. Morse, R.A.; Calderone, N.W. The value of honey bees as pollinators of U.S. crops in 2000. *Bee Cult. Mag.* **2000**, *128*, 1–15.
4. Cârnu, I. *Plante Melifere*; Editura Ceres: Bucharest, Romania, 1972.
5. Celli, G.; Maccagnani, B. Honey bees as bioindicators of environmental pollution. *Bull. Insectology* **2003**, *56*, 137–139.
6. Negri, I.; Mavris, C.; Di Prisco, G.; Caprio, E.; Pellecchia, M. Honey bees (*Apis mellifera*, L.) as active samplers of airborne particulate matter. *PLoS ONE* **2015**, *10*, e0132491. [CrossRef] [PubMed]
7. Bargańska, Ż.; Ślebioda, M.; Namieśnik, J. Honey bees and their products: Bioindicators of environmental contamination. *Crit. Rev. Environ. Sci. Technol.* **2016**, *46*, 235–248. [CrossRef]
8. Barmaz, S.; Potts, S.G.; Vighi, M. A novel method for assessing risks to pollinators from plant protection products using honey bees as a model species. *Ecotoxicology* **2010**, *19*, 1347–1359. [CrossRef]
9. Bromenshenk, J.J.; Henderson, C.B.; Seccomb, R.A.; Welch, P.M.; Debnam, S.E.; Firth, D.R. Bees as biosensors: Chemosensory ability, honey bee monitoring systems, and emergent sensor technologies derived from the pollinator syndrome. *Biosensors* **2015**, *5*, 678–711. [CrossRef]
10. Arcega-Rustia, D.J.; Ngo, T.N.; Ta-Te, L. An IoT based information system for honeybees in and out activity with beehive environmental condition monitoring. In Proceedings of the International Symposium on Machinery and Mechatronics for Agricultural and Biosystems Engineering 2016, Niigata, Japan, 23–25 May 2016; Available online: https://www.researchgate.net/publication/320134391 (accessed on 2 November 2019).
11. Debauche, O.; El Moulat, M.; Mahmoudi, S.; Boukraa, S.; Manneback, P.; Lebeau, F. Web monitoring of bee health for researchers and beekeepers based on the internet of things. *Procedia Comput. Sci.* **2018**, *130*, 991–998. [CrossRef]
12. Gil-Lebrero, S.; Quiles-Latorre, F.J.; Ortiz-López, M.; Sánchez-Ruiz, V.; Gámiz-López, V.; Luna-Rodríguez, J.J. Honey bee colonies remote monitoring system. *Sensors* **2017**, *17*, 55. [CrossRef]
13. Meikle, W.G.; Holst, N. Application of continuous monitoring of honeybee colonies. *Apidologie* **2015**, *46*, 10–22. [CrossRef]
14. Pesovic, U.; Randic, S.; Stamenkovic, Z. Design and implementation of hardware platform for monitoring honeybee activity. In Proceedings of the 4th International Conference on Electrical, Electronics and Computing Engineering, IcETRAN 2017, Kladovo, Serbia, 5–8 June 2017.
15. Siceanu, A.; Căuia, E.; Rădoi, C.; Gureșoaie, I.; Svasta, P.; Vulpe, V.; Davidescu, M.; Ionescu, C. Electronic equipment to monitorize some biological process of economic importance in honeybee colony and its environment. *Sci. Pap. Anim. Sci. Biotechnol.* **2007**, *40*, 134–140.
16. Siceanu, A. Sistemul de stup intelligent Simfony (The intelligent hive Simfony system). *Rom. Apic.* **2017**, *5*, 24–25.
17. Sánchez-Bayo, F.; Wyckhuys, K.A.G. Worldwide decline of the entomofauna: A review of its drivers. *Biol. Conserv.* **2019**, *232*, 8–27. [CrossRef]
18. Gray, A.; Brodschneider, R.; Adjlane, N.; Ballis, A.; Brusbardis, V.; Charrière, J.-D.; Chlebo, R.; Coffey, M.F.; Cornelissen, B.; Amaro da Costa, C.; et al. Loss rates of honey bee colonies during winter 2017/18 in 36 countries participating in the COLOSS survey, including effects of forage sources. *J. Apic. Res.* **2019**. [CrossRef]
19. Potts, S.G.; Roberts, S.P.M.; Dean, R.; Marris, G.; Brown, M.A.; Jones, R.; Neumann, P.; Settele, J. Declines of managed honey bees and beekeepers in Europe. *J. Apic. Res.* **2010**, *49*, 15–22. [CrossRef]

20. Sanchez-Bayo, F.; Goka, K. Pesticide Residues and Bees—A Risk Assessment. *PLoS ONE* **2014**, *9*, e94482. [CrossRef]

21. Blacquiere, T.; Smagghe, G.; van Gestel, C.M.; Mommaerts, V. Neonicotinoids in bees: A review on concentrations, side-effects and risk assessment. *Ecotoxicology* **2012**, *21*, 973–992. [CrossRef]

22. Chauzat, M.-P.; Martel, A.-C.; Cougoule, N.; Porta, P.; Lachaize, J.; Zegane, S.; Aubert, M.; Carpentier, P.; Faucon, J.-P. An assessment of honey bee colony matrices, *Apis mellifera* (Hymenoptera: *Apidae*) to monitor pesticide presence in continental France. *Environ. Toxicol. Chem.* **2011**, *30*, 103–111. [CrossRef]

23. Mitchell, E.A.D.; Mulhauser, B.; Mulot, M.; Mutabazi, A.; Glauser, G.; Aebi, A. A worldwide survey of neonicotinoids in honey. *Science* **2017**, *358*, 109–111. [CrossRef]

24. Botías, C.; David, A.; Horwood, J.; Abdul-Sada, A.; Nicholls, E.; Hill, E.; Goulson, D. Neonicotinoid residues in wildflowers, a potential route of chronic exposure for bees. *Environ. Sci. Technol.* **2015**. [CrossRef] [PubMed]

25. Böhme, F.; Bischoff, G.; Zebitz, C.P.W.; Rosenkranz, P.; Wallner, K. Pesticide residue survey of pollen loads collected by honeybees (*Apis mellifera*) in daily intervals at three agricultural sites in South Germany. *PLoS ONE* **2018**, *13*, e0199995. [CrossRef] [PubMed]

26. Bonmatin, J.M.; Giorio, C.; Girolami, V.; Goulson, D.; Kreutzweiser, D.P.; Krupke, C.; Liess, M.; Long, E.; Marzaro, M.; Mitchell, E.A.D.; et al. Environmental fate and exposure; neonicotinoids and fipronil. *Environ. Sci. Pollut. Res.* **2015**. [CrossRef] [PubMed]

27. Bonmatin, J.M.; Moineau, R.; Charvet, C.; Fleche, M.E.; Colin, E.; Bengsch, R. A LC/APCI-MS/MS Method for analysis of imidacloprid in soils, in plants, and in pollens. *Anal. Chem.* **2003**, *75*, 2027–2033. [CrossRef]

28. Rortais, A.; Arnold, G.; Halm, M.-P.; Touffet-Briens, F. Modes of honeybees exposure to systemic insecticides: Estimated amounts of contaminated pollen and nectar consumed by different categories of bees. *Apidologie* **2005**. [CrossRef]

29. Samson-Robert, O.; Labrie, G.; Chagnon, M.; Fournier, V. Planting of neonicotinoid-coated corn raises honey bee mortality and sets back colony development. *PeerJ* **2017**, *5*, e3670. [CrossRef]

30. Guez, D.; Suchail, S.; Gauthier, M.; Maleszka, R.; Belzunces, L.P. Contrasting effects of imidacloprid on habituation in 7- and 8-day-old honeybees (*Apis mellifera*). *Neurobiol. Learn. Mem.* **2001**, *76*, 183–191. [CrossRef]

31. European Food Safety Authority. Evaluation of the data on clothianidin, imidacloprid and thiamethoxam for the updated risk assessment to bees for seed treatments and granules in the EU. *EFSA Support. Publ.* **2018**. [CrossRef]

32. Bonmatin, J.M.; Marchand, P.A.; Charvet, R.; Moineau, I.; Bengsch, E.R.; Colin., M.E. Quantification of imidacloprid uptake in maize crops. *J. Agric. Food Chem.* **2005**, *53*, 5336–5341. [CrossRef]

33. Siceanu, A.; Căuia, E. Analize de neonicotinoide în România—Rezultatele parțiale obținute în urma derulării unui proiect de cercetare (Analyses of neonicotinoids in Romania—The preliminary results obtained following a research project). *Rom. Apic.* **2018**, *10*, 10–16.

34. Siceanu, A.; Căuia, E. Analize de neonicotinoide în România—Rezultate finale si concluzii (Analyses of neonicotinoids in Romania—The final results and conclusions). *Rom. Apic.* **2018**, *11*, 3–9.

35. Tsvetkov, N.; Samson-Robert, O.; Sood, K.; Patel, H.S.; Malena, D.A.; Gajiwala, P.H.; Maciukiewicz, P.; Fournier, V.; Zayed, A. Chronic exposure to neonicotinoids reduces honey-bee health near corn crops. *Science* **2017**. [CrossRef] [PubMed]

36. Sánchez-Hernández, L.; Hernández-Domínguez, D.; Martín, M.T.; Nozal, M.J.; Higes, M.; Bernal Yagüe, J.L. Residues of neonicotinoids and their metabolites in honey and pollen from sunflower and maize seed dressing crops. *J. Chromatogr. A* **2016**, *1428*, 220–227. [CrossRef]

37. European Food Safety Authority. Statement on the assessment of the scientific information from the Italian project "APENET" investigating effects on honeybees of coated maize seeds with some neonicotinoids and fipronil. *EFSA J.* **2012**, *10*, 2792. [CrossRef]

38. Online Petition for Honeybee Colones Losses in Romania: 2015. Available online: https://www.petitieonline.com/sos_albina_romaneasc (accessed on 25 July 2019).

39. The European Academies' Science Advisory Council. *Ecosystem Services, Agriculture and Neonicotinoids*; EASAC Policy Report; German National Academy of Sciences Leopoldina: Halle (Saale), Germany, 2015; ISBN 978-3-8047-3437-1. Available online: www.easac.eu (accessed on 27 November 2017).

40. Suchail, S.; Guez, D.; Belzunces, L. Discrepancy between acute and chronic toxicity induced by imidacloprid and its metabolites in *Apis mellifera*. *Environ. Toxicol. Chem.* **2001**, *20*, 2482–2486. [CrossRef] [PubMed]

41. Tomizawa, M.; Casida, J.E. Neonicotinoid insecticide toxicology: Mechanisms of selective action. *Annu. Rev. Pharm. Toxicol.* **2005**, *45*, 247–268. [CrossRef] [PubMed]

42. Yamamoto, I.; Yabuta, G.; Tomizawa, M.; Saito, T.; Miyamoto, T.; Kagabu, S. Molecular mechanism for selective toxicity of nicotinoids and neonicotinoids. *J. Pestic. Sci.* **1995**, *20*, 33–40. [CrossRef]

43. Suchail, S.; Guez, D.; Belzunces, L. Characteristics of imidacloprid toxicity in two *Apis mellifera* subspecies. *Env. Toxicol. Chem.* **2000**, *19*, 1901–1905. [CrossRef]

44. Rondeau., G.; Sánchez-Bayo, F.; Tennekes, H.A.; Decourtye, A.; Ramírez-Romero, R.; Desneux, N. Delayed and time-cumulative toxicity of imidacloprid in bees, ants and termites. *Sci. Rep.* **2014**, *4*, 5566. [CrossRef]

45. Simon-Delso, N.; Amaral-Rogers, V.; Belzunces, L.P.; Bonmatin, J.M.; Chagnon, M.; Downs, C.; Furlan, L.; Gibbons, D.W.; Giorio, C.; Girolami, V.; et al. Systemic insecticides (neonicotinoids and fipronil): Trends, uses, mode of action and metabolites. *Environ. Sci. Pollut. Res.* **2015**, *22*, 5–34. [CrossRef]

46. Bortolotti, L.; Montanari, R.; Marcelino, J.; Medrzycki, P.; Maini, S.; Porrini, C. Effects of sub-lethal imidacloprid doses on the homing rate and foraging activity of honey bees. *Bull. Insectol.* **2003**, *56*, 63–67.

47. Decourtye, A.; Armengaud, C.; Renou, M.; Devillers, J.; Cluzeau, S.; Gauthier, M.; Pham-Delegue, M.H. Imidacloprid impairs memory and brain metabolism in the honeybee (*Apis Mellifera* L.). *Pestic. Biochem. Phys.* **2004**, *78*, 83–92. [CrossRef]

48. Fischer, J.; Muller, T.; Spatz, A.-K.; Greggers, U.; Grunewald, B.; Menzel, R. Neonicotinoids interfere with specific components of navigation in honeybees. *PLoS ONE* **2014**, *9*, e91364. [CrossRef] [PubMed]

49. Tan, K.; Chen, W.; Dong, S.; Liu, X.; Wang, Y.; Nieh, J.C. Imidacloprid alters foraging and decreases bee avoidance of predators. *PLoS ONE* **2014**, *9*, e102725. [CrossRef] [PubMed]

50. Tosi, S.; Burgio, G.; Nieh, J.C. A common neonicotinoid pesticide, thiamethoxam, impairs honey bee flight ability. *Nat. Sci. Rep.* **2017**, *7*, 1201. [CrossRef]

51. Yang, E.C.; Chang, H.C.; Wu, W.Y.; Chen, Y.W. Impaired olfactory associative behavior of honeybee workers due to contamination of imidacloprid in the larval stage. *PLoS ONE* **2012**, *7*, e49472. [CrossRef]

52. Baines, D.; Wilton, E.; Pawluk, A.; Gorter, M.; Chomistek, N. Neonicotinoids act like endocrine disrupting chemicals in newly-emerged bees and winter bees. *Nat. Sci. Rep.* **2017**, *7*, 10979. [CrossRef]

53. Hatjina, F.; Papaefthimiou, C.; Charistos, L.; Dogaroglu, T.; Bouga, M.; Emmanouil, C.; Arnold, G. Sublethal doses of imidacloprid decreased size of hypopharyngeal glands and respiratory rhythm of honeybees in vivo. *Apidologie* **2013**. [CrossRef]

54. Wu-Smart, J.; Spivak, M. Sub-lethal effects of dietary neonicotinoid insecticide exposure on honey bee queen fecundity and colony development, www.nature.com/scientificreports. *Sci. Rep.* **2016**, *6*, 32108. [CrossRef]

55. Straub, L.; Villamar-Bouza, L.; Bruckner, S.; Chantawannakul, P.; Gauthier, L.; Khongphinitbunjong, K.; Retschnig, G.; Troxler, A.; Vidondo, B.; Neumann, P.; et al. Neonicotinoid insecticides can serve as inadvertent insect contraceptives. *Proc. R. Soc. B Biol. Sci.* **2016**. [CrossRef]

56. Lu, C.; Warchol, K.M.; Callahan, R.A. Sub-lethal exposure to neonicotinoids impaired honey bees, winterization before proceeding to colony collapse disorder. *Bull. Insectol.* **2014**, *67*, 125–130.

57. Meikle, W.G.; Adamczyk, J.J.; Weiss, M.; Gregorc, A. Effects of bee density and sublethal imidacloprid exposure on cluster temperatures of caged honey bees. *Apidologie* **2018**, *49*, 581–593. [CrossRef]

58. Alaux, C.; Brunet, J.-L.; Dussaubat, C.; Mondet, F.; Tchamitchan, S.; Cousin, M.; Brillard, J.; Baldy, A.; Belzunces, L.P.; Le Conte, Y. Interactions between *Nosema* microspores and a neonicotinoid weaken honeybees (*Apis mellifera*). *Environ. Microbiol.* **2009**. [CrossRef]

59. Aufauvre, J.; Biron, D.G.; Vidau, C.; Fontbonne, R.; Roude, M.; Diogon, M.; Viguès, B.; Belzunces, L.P.; Delbac, F.; Blot, N. Parasite-insecticide interactions: A case study of *Nosema ceranae* and fipronil synergy on honeybee. *Nat. Sci. Rep.* **2012**, *2*, 326. [CrossRef] [PubMed]

60. Hernández, L.J.; Krainer, S.; Engert, A.; Schuehly, W.; Riessberger-Gallé, U.; Crailsheim, K. Sublethal pesticide doses negatively affect survival and the cellular responses in American foulbrood-infected honeybee larvae. *Sci. Rep.* **2017**, *7*, 40853. [CrossRef]

61. Pettis, J.S.; VanEngelsdorp, D.; Johnson, J.; Dively, G. Pesticide exposure in honey bees results in increased levels of the gut pathogen *Nosema*. *Naturwissenschaften* **2012**, *99*, 153–158. [CrossRef]

62. Vidau, C.; Diogon, M.; Aufauvre, J.; Fontbonne, R.; Viguès, B.; Brunet, J.-L.; Texier, C.; Biron, D.G.; Blot, N.; El Alaoui, H.; et al. Exposure to sublethal doses of fipronil and thiacloprid highly increases mortality of honeybees previously infected by Nosema Ceranae. *PLoS ONE* **2011**, *6*, e21550. [CrossRef]
63. Van der Sluijs, J.P.; Simon-Delso, N.; Goulson, D.; Maxim, L.; Bonmatin, J.-M.; Belzunces, L.P. Neonicotinoids, bee disorders and the sustainability of pollinator services. *Curr. Opin. Environ. Sustain.* **2013**. [CrossRef]
64. Di Prisco, G.; Cavaliere, V.; Annoscia, D.; Varricchio, P.; Caprio, E.; Nazzi, F.; Gargiulo, G.; Pennacchio, F. Neonicotinoid clothianidin adversely affects insect immunity and promotes replication of a viral pathogen in honey bees. *Proc. Natl. Acad. Sci. USA* **2013**, *110*, 18466–18471. [CrossRef]
65. European Food Safety Authority. Conclusion on the peer review of the pesticide risk assessment of the active substance thiacloprid. *EFSA J.* **2019**, *17*, 5595. [CrossRef]

 *diversity*

Article

Hydroxymethylfurfural Affects Caged Honey Bees (*Apis mellifera carnica*)

Aleš Gregorc [1],*, Snežana Jurišić [1] and Blair Sampson [2]

[1] Faculty of Agriculture and Life Sciences, University of Maribor, Pivola 10, 2311 Hoče, Slovenia; ssjurisic@gmail.com

[2] USDA-ARS Thad Cochran Southern Horticultural Laboratory, Poplarville, MS 39470, USA; Blair.Sampson@ars.usda.gov

* Correspondence: ales.gregorc@um.si

Received: 28 November 2019; Accepted: 28 December 2019; Published: 31 December 2019

Abstract: A high concentration of hydroxymethylfurfural (HMF) (e.g., 15 mg HMF per kg honey) indicates quality deterioration for a wide range of foods. In honey bee colonies, HMF in stored honey can negatively affect bee health and survival. Therefore, in the laboratory, we experimentally determined the effects of HMF on the longevity and midgut integrity of worker *Apis mellifera carnica* by feeding bees standard diets containing five concentrations of HMF (100, 500, 1000, and 1500 ppm). Simultaneously, we also examined HMF's effect on *Nosema ceranae* spore counts within infected honey bees. We performed an immunohistochemical analysis of the honey bee midgut to determine possible changes at the cellular level. No correlation was established between HMF concentration and *N. ceranae* spore counts. Negative effects of HMF on bees were not observed in the first 15 days of exposure. However, after 15 to 30 days of exposure, HMF caused midgut cells to die and an increased mortality of honey bee workers across treatment groups.

Keywords: hydroxymethylfurfural; honey bee; cell death; immunohistochemistry; *Nosema ceranae*

---

1. Introduction

Honey is a highly concentrated mixture of mainly two dissolved sugars, fructose and glucose, plus at least 22 other composite sugars [1] and 70 other compounds including proteins, vitamins, minerals, organic acids, aromatic compounds, and various derivatives of chlorophyil [2]. Many more honey components may remain undiscovered. Therefore, identifying potential fraudulent honeys by component analyses may be difficult unless specific breakdown products (metabolites) can be identified such as substances like hydroxymethylfurfural, or 5-hydroxymethyl-2-furaldehyde ($C_6H_6O_3$) (HMF). HMF can be selectively produced from keto-hexose—notably from D-fructose [3] and other acidic media containing dissolved monosaccharides [4]. Normally, HMF is present in honey in trace amounts [5]. The rate of HMF formation in foods depends on environmental temperature, the type of sugar, pH, and the concentration of divalent cations in the medium [6,7]. Excessive heating or inappropriate storage conditions can increase HMF levels, which are recognized as a marker of quality deterioration for a wide range of foods containing carbohydrates [8]. Inappropriate heat processing of honey affects honey fermentation and reduces honey quality [9]. In fresh honey, HMF can occur at concentrations as high as 15 mg HMF/kg, but it normally occurs at levels between 0.06–0.2 mg HMF/kg [5]. For the most part, HMF is naturally present in honey, and at low concentrations (e.g., ~100–500 ppm) it does not reduce honey quality; it could thus be used as an identifier of a honey's origin and quality. The Codex Alimentarius of the World Health Organization (WHO) and the European Union (EU Directive 110/2001) have defined a maximum HMF quantity level in heat-treated honeys (40 mg HMF/kg) above which honey quality begins to deteriorate. HMF concentration increases above 20 °C. Temperatures inside a hive normally exceed 20 °C (~28–30 °C) and in summer can reach as high as 40 °C or more, when the concentration of HMF can reach 10 mg/kg of honey [10], a level one-third of that known to be harmless to bees (30 mg/kg) [11]. Although these high levels of HMF are considered nontoxic to bees, few studies actually confirm a safe level of HMF in honey bee colonies [12]. Concentrations of HMF <10–15 mg/kg in honey pose little risk to honey bees, but toxic concentrations of HMF seems to induce lethal intestinal tract ulceration [12]. About 150 mg HMF/kg of commercially acid-hydrolyzed inverted sugar syrup can cause 50% bee mortality within 16 days [11]. The HMF concentration in inverted syrup for feeding bees should not exceed 20 mg/kg, as in most honeys [13].

There is no standard limit value of HMF in bee nourishment. Approximately 250 ppm HMF in the honey bee diet is considered toxic [14]. High concentrations of HMF in stored honey could represent a factor in the early death of bees and in the extinction of honey bee colonies [15]. It is therefore important to understand the potential adverse effects of high HMF doses on honey bees. Thus, the objective of this study was to determine the toxic effects and mode of action of HMF on caged bees fed in laboratory assays. We also used an immunohistochemical assay to examine the impact of HMF toxicity on the cellular death of epithelial cells lining the worker midgut.

2. Materials and Methods

2.1. Toxicological Tests

Toxicological tests on the autochthonous Carniolan honey bee *Apis mellifera carnica* occurred in laboratory conditions. Worker bees were maintained in incubators set at a near-constant 28 °C and at ~65% relative humidity (RH), according to normal temperatures outside the brood nest [16,17]. This was to simulate the environment outside the brood nest, where we normally install sugar patties in honey bee colonies.

Plastic ~8 cm (H) × ~12 cm (Dia.) "cake" boxes originally designed to hold up to 100 CDs were repurposed as bee containment units (test chambers) by drilling ~80 circular ventilation holes, each ~2 mm wide, into the top cover. Similar experimental plastic box approaches have been used in previous experiments [18–20]. Two additional holes of 12 mm diameter were added as placeholders for plastic feeding tubes, which provided bees with Apifonda sugar candy and water, respectively. Apifonda contains sucrose, glucose syrup, and invert sugar syrup and corresponds to the nutritional value of 0.9 kg crystalline sugar (Südzucker Sugar, Mannheim, Germany). To simulate a colony habitat, each containment unit contained a 4 cm × 5 cm piece of wax foundation. We divided the bees into five groups. Each group of 70 bees was placed into its own individual containment unit. We provided thoroughly homogenized Apifonda as bee nourishment, into which was added HMF (5-hydroxymethyl-2-furaldehyde, Sigma Aldrich) at concentrations of 0 mg HMF/kg candy (control), 100 mg HMF/kg, 500 mg HMF/kg, 1000 mg HMF/kg, and 1500 mg HMF/kg. We took the capped brood from three clinically healthy honey bee colonies a day before starting the experiment and placed combs into the incubator at 34.5 °C and 65% RH. The next day, newly emerged 0–24 h old bees were placed into the containment units. Each HMF-treated group was replicated five times. We recorded the daily food consumption and daily mortality rates of the confined worker bees.

Dead bees exposed to the control and HMF concentrations were counted and numbers were deducted from the initial bee population ($n = 70$). The survival rate was calculated as 100% minus mortality per control or treatment groups. Data analyses were performed using ANOVA (analysis of variance) in Statgraphic [21]. Mean bee survival rates were compared among the treatment groups with a one-way ANOVA and mean separation was accomplished with Tukey tests.

2.2. Immunohistochemical Analyses

We established 5 groups of 50 bees. Brood combs were obtained from clinically healthy honey bee colonies in the same way as described for the toxicological tests. Newly emerged 0–24 h old bees were placed into the containment units. Bees in the first group received no HMF added to the Apifonda candy. Group 2 received Apifonda containing 100 mg/kg of HMF. Group 3 received

Apifonda containing 500 mg/kg of HMF. Group 4 received Apifonda containing 1000 mg/kg of HMF. Group 5 received Apifonda containing 1500 mg /kg of HMF. Three bees from each of these treatment groups were randomly sampled at 5 day intervals: on day 5, 10, 15, and 20. Sampled bees were anesthetized by subjecting them to cold for about 10 min, and their midgut was removed. Midguts were fixed in 10% neutral buffered formalin, dehydrated in ethanol, and embedded in paraffin wax, which was sliced into 5 μm sections that were then de-paraffinized and processed following the instructions provided with the In Situ Cell Death Detection Kit AP' (ISCDDK) (Roche, Mannheim, Germany). The EnVision System alkaline phosphatase kit (Dako) was used to obtain a red-colored precipitate in the sections treated with the ISCDDK assay reagents. The sections were counterstained with hematoxylin. TUNEL-positive cells possessed red nuclei, which indicate the reaction products of cell death. TUNEL-negative nuclei of healthy, intact cells appeared blue. A control labeling of midgut tissue was accomplished by substituting the deoxynucleotidyl transferase (TdT) enzyme with phosphate-buffered saline (PBS) during the TUNEL reaction. Sections were mounted in Faramount aqueous mounting medium (Dako). Slide contents were analyzed and digitally photo-documented with a bright field light microscope at 400× magnification. We repeated this immunohistochemical experiment twice.

2.3. Semi-Quantitative Analysis of Cell Death

TUNEL-labeled tissue slides were used for the quantification of cell type and cell death using ISCDDK. For each experimental group of bees, approximately 300 total cells from each individual (three bees at different collection times per group) were counted in random fields on different slides. The results were expressed as the proportion of cells with positive staining. To confirm reproducibility, 25% of the slides were chosen randomly and scored twice [22,23].

2.4. Nosema Ceranae Spore Counts

To quantify *N. ceranae* infection in sampled bees, we temporarily stored dead bees from separate containment units in a freezer. All dead bees in consecutive days from experimental cages were sampled and examined for *N. ceranae* spores. Nosema spores on adult bees were potentially derived from combs where bees emerged; we assumed that spores were equally distributed between bees. Altogether, 350 bees were examined microscopically for *N. cerane* spores across all treatment groups. Species determination of *N. ceranae* in these frozen bee samples was confirmed using multiplex PCR [24]. Spore counts were made using a Bürker hemocytometer and a Zeiss light microscope under phase contrast (Axioskop 2 Plus, Zeiss, Jena, Germany). Spore samples were extracted from a bee's detached abdomen by grinding the tissue in 1 mL water with a mortar and pestle. A drop (~60 μL) of this homogenous suspension was applied to a hemocytometer and counting was conducted a few minutes later when spores fully settled. The average number of *N. ceranae* spores was calculated after counting the number of spores in all four outer squares divided by four for each dissected bee.

3. Results

3.1. Longevity of Bees

About 90% of caged bees in each treatment group survived the first 15 days (Figure 1). After day 30, less than 30% of bees were left alive in the five treatment groups, which included the Apifonda control. The steepest declines in bee survival occurred between days 15–30 post-treatment (Figure 1). Treatment differences were detected among the five treatment groups (Figure 1; F = 2.968; df = 4; p = 0.028). The longevity of the bees receiving the highest HMF concentration (1500 mg HMF/kg) was reduced when compared with other HMF treatment groups and untreated control bees (Figure 2).

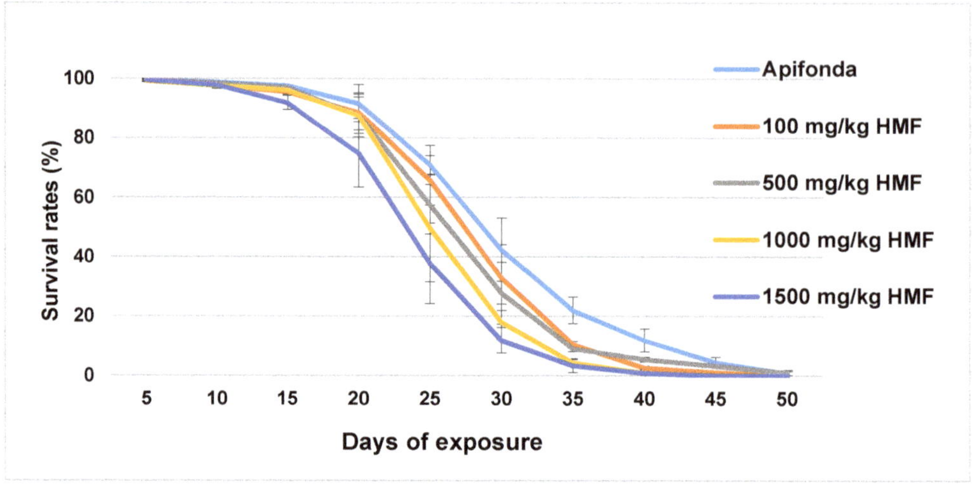

Figure 1. Survival rates of caged bees (expressed in %) are shown in 5 day intervals for each treatment group and the untreated control group (Apifonda). Bars indicate ± SD.

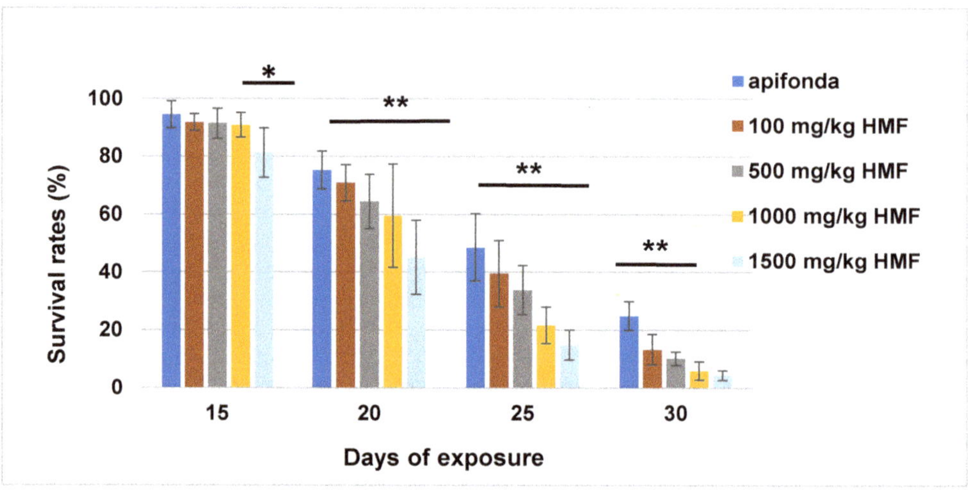

Figure 2. The relative number of live caged bees on the 15th, 20th, 25th, and 30th day of the experiment. The figure shows dates with the highest differences of survived bees between treatment groups. Asterisks indicate statistically significant differences between treatment groups. The Tukey test shows a difference at $p < 0.05$ (*) and at $p < 0.01$ (**).

3.2. Nosema Ceranae Spores

After newly emerged bees were fed on 100 mg/kg HMF for 10 days, *Nosema* was detected in dead bees at an average density of 2.6×10^6 spores/bee. From day 11 until the end of the experiment, spores were present in variable numbers in each of the treatment groups. The highest number of spores (6.6×10^6) was observed between days 26 and 30 in untreated control bees (Figure 3). From day 21 until the end of experiment, the highest number of spores per individual bee (5.76×10^6) was noted in the group receiving 100 mg/kg of HMF in Apifonda candy. There were no statistically significant differences between the groups in the interval between day 10 and day 20 (F = 0.642; df = 4; $p = 0.634$). From day 21 until the end of the experiment, there was a statistically significant difference between treatment groups. A high density of *Nosema* spores (3.55×10^6) was detected in the youngest bees (10–15 days) exposed to 1500 mg/kg of HMF in Apifonda, and a lower *N. ceranae* spore density was found in bees that died later in the course of the experiment.

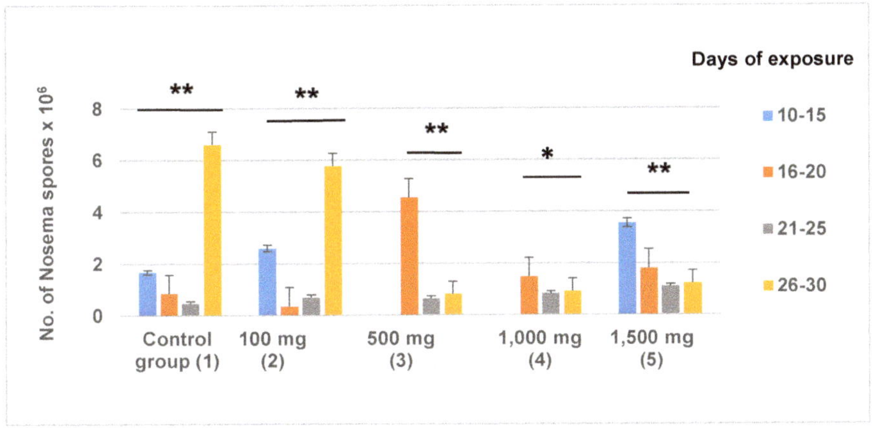

Figure 3. The number of *Nosema ceranae* spores counted in dissected bees. The average values are expressed in millions of spores per worker bee. HMF quantity is expressed as mg/kg of HMF in Apifonda candy. The control group received only Apifonda candy. Asterisks indicate a statistically significant difference of *N. ceranae* spore counts inside a treatment group on different sampling dates. The Tukey test shows a difference at $p < 0.05$ (*) and at $p < 0.01$ (**). Note that *N. ceranae* spores were not detected in bees from treatment groups 3 and 4 at the first sampling date. Across all treatment group, 350 bees were examined.

3.3. Immunohistochemical Analyses

Varieties of apoptotic midgut cell deletions were observed in bees exposed to different HMF concentrations for 10 days. The amount of immunohistochemically positive cells was not dependent on HMF concentration between 5 and 30 days (Figure 4). However, in bees exposed to the lowest dose of HMF, cell death was observed in the epithelium of the midgut (sporadic positive cells) but at a level below that of bees fed higher concentrations of HMF. In contrast, the level of cell death in the midgut epithelium of bees fed with candy containing 500 mg/kg of HMF in Apifonda remained high throughout the bioassay (Table 1).

Table 1. Results of the semi-qualitative analyses of midgut cell death using the In Situ Cell Death Detection Kit (ISCDDK) immunohistochemical method. The results represent the percentage of positive cells in midgut tissue treated with ISDDK: (I) individual positive cells; (II) range between 0% and 20% positive cells; (III) range between 21% and 60% positive cells; (IV) range between 61% and 90% positive cells; (x) no samples available. Treatment groups: 1. Untreated control; 2. 100 mg/kg of HMF in Apifonda; 3. 500 mg/kg of HMF in Apifonda; 4. 1000 mg/kg of HMF in Apifonda; 5. 1500 mg/kg of HMF in Apifonda.

Age of Bees (Days)	Treatment Groups				
	1 (control)	2 (100 mg HMF)	3 (500 mg HMF)	4 (1000 mg HMF)	5 (1500 mg HMF)
5	I	I	I	IV	IV
10	II	I	IV	III	I
15	II	III	IV	IV	IV
20	IV	I	IV	II	III
25	I	I	IV	III	I
30	x	x	III	I	I

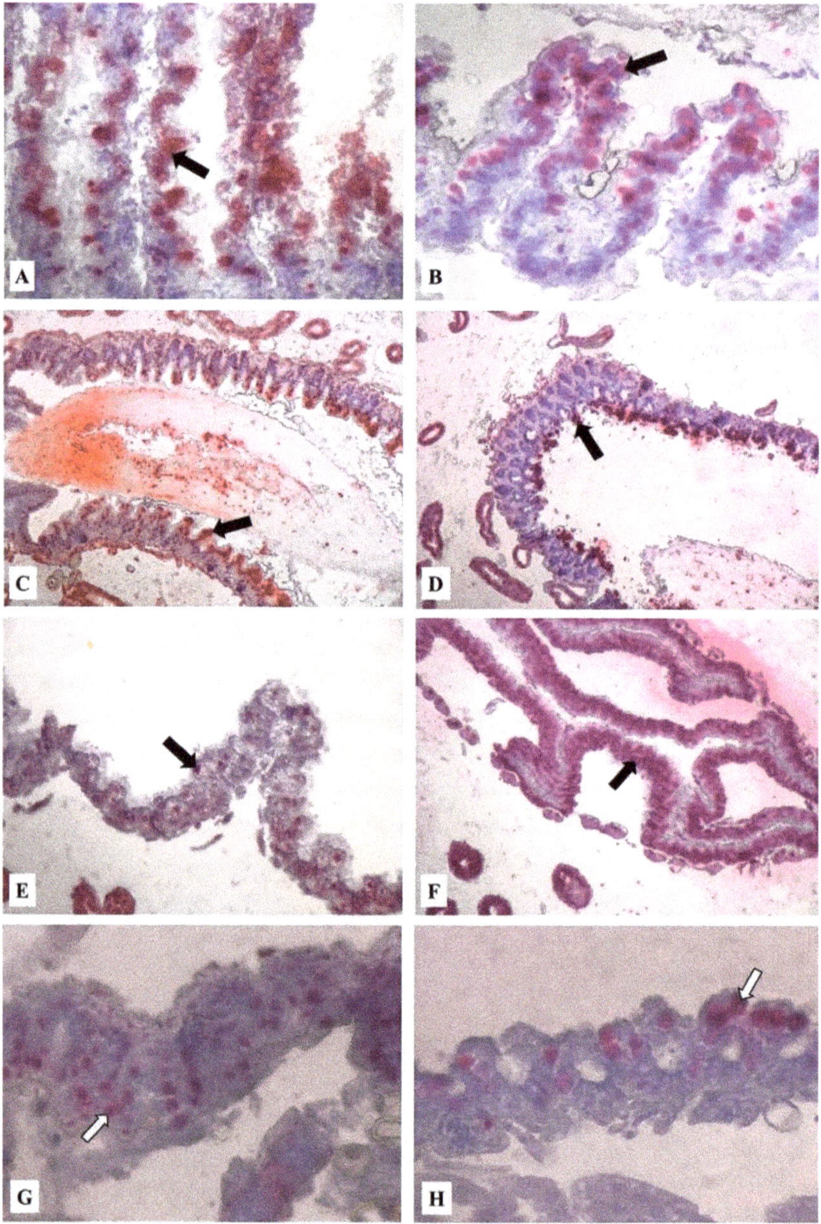

Figure 4. Midgut of formalin-fixed, paraffin-embedded tissue of worker bees exposed to HMF. Detection of programmed cell death by TUNEL using the In Situ Cell Death Detection Kit. Red azo-dye staining is localized in midgut epithelial cell nuclei (arrows). (**A**) 100 mg/kg of HMF in Apifonda diet (5 days), magnification 200×; (**B**) 500 mg/kg of HMF (5 days), magnification 200×; (**C**) 100 mg/kg of HMF (10 days), magnification 100×; (**D**) 1500 mg/kg of HMF (10 days), magnification 100×; (**E**) control group (15 days), magnification 200×. (**F**) 1000 mg/kg of HMF (15 days), magnification 200×; (**G**) 500 mg/kg of HMF (20 days), magnification 400×; (**H**) control group (5 days), magnification 400×. The dosage of HMF was incorporated to 1 kg of Apifonda candy.

Notably, the proportion of positive midgut digestive cells in the bees treated with 100 and 500 mg/kg HMF in Apifonda (Figure 4A,B) was much lower than in the groups exposed to 1000 or 1500'mg/kg HMF. In bees exposed to 100 mg/kg of HMF, affected midgut epithelial cells were vacuolated with apical cell fragments released into the lumen (Figure 4A). Similar vacuolization was observed in bees exposed to 500 mg/kg HMF in Apifonda (Figure 4B). Midgut cells with positive red reaction products in the epithelium apical region were also seen in bees fed with candy containing 100 or 1500 mg/kg of HMF for 10 days (Figure 4C,D) and also in bees fed with candy containing 1500 mg/kg of HMF for 15 days (Figure 4F). Midgut epithelial cells were still intact and attached to the basal membrane even at the highest HMF concentration over the course of the experiment. In the control group of untreated bees (without added HMF), sporadic midgut cells with specific red reaction products were found (Figure 3EH). In live bees exposed to any HMF dose for 25 or 30 days, no morphological alterations or increased levels of positive ISCDDK cells were seen (Figure 4G). Bees fed with the two highest HMF concentrations, 1000 or 1500 mg/kg of HMF, and surviving to 25–30 days displayed few midgut cells with positive reaction products, an observation similar to the reaction levels displayed by untreated control bees. Taken together, the highest levels of ISCDDK-positive midgut epithelial cells were found in bees fed with 500, 1000, or 1500 mg/kg of HMF in Apifonda from 15 to 20 days. After that time, there was a notable decrease in the proportion of midgut cells with specific red reaction products. The most persistent and high levels of apoptotic cells throughout the experiment were found in bees fed with 500 mg/kg of HMF. The proportion of apoptotic cells remained high when compared with all treatment groups, including the controls. We did not include *Nosema*-infested bees in ISCDDK assays in order to exclude the effect of parasites on midgut cell death. The lowest HMF dose effect at the first and second sampling dates (5 and 10 days) was demonstrated by red reaction products in midgut epithelial cells (Figure 4A,C).

4. Discussion

Carniolan honey bees exposed to HMF suffer increased death, especially after feeding on it for 15 to 30 days. Bee survival, therefore, could be improved by reducing worker exposure to HMF in their foodstuffs. In a large-scale experiment, a relatively low concentration of HMF (150 mg/kg) found in acid-hydrolyzed inverted sugar syrup induced 50% bee mortality within 16 days after the start of feeding. Likewise, high-fructose corn syrup, a saccharose replacement for honey bees, containing 150 mg/kg HMF also induces 50% of bee mortality within about 19 days of feeding [14]. An HMF concentration of 30 mg/kg to 48 mg/kg is supposed to be harmless to honey bee workers, whether or not bees are overwintering [11]. The concentration of HMF in winter food stores is lower than that of initial syrups deposited by bees. Perhaps honey bees can safely metabolize or detoxify small quantities of HMF in their stored foods [25]. Except for these small-scale studies, few studies have systematically examined the effect of HMF on bee health and physiology, which is surprising given that concentrations of 7500 ppm HMF (mg/kg) or higher have been reported to cause massive bee kills, i.e., 100% mortality of bee larvae with LC_{50} for larvae ranging from 2424 ppm to 4280 ppm. Thus, larvae appear far more sensitive to HMF than adult honey bees [26]. Clearly, our results coupled with evidence cited earlier confirm the toxicity of HMF to adult and developing honey bees and the need for its mitigation in commercial bee yards. A deadly chemical interaction may occur when HMF builds up in particular foodstuffs. For instance, sugar syrup or high-fructose corn syrup enhance the acute toxicity of HMF and double rates of honey bee mortality (from 40% in Apifonda controls to 80% in sugar syrups) at a high concentration of around 1500 mg/kg of HMF when compared with its combination with Apifonda candy. The effects of HMF in the first two weeks of feeding are sublethal for bees, but after 14 days, HMF starts to become fatal. More studies are needed to assess the tolerances of honey bees to prolonged oral exposure to HMF.

The mechanism through which HMF sickens or kills honey bees is unknown. We surmise that the midgut is the first bodily tissue exposed to the activity of HMF as well as to other toxins, including pathogens. Midgut epithelial cells are the first line of defense against both pathogens and toxins [27]; [28] and potential dysbiosis together with other disorders can present an important source of energy dysregulation [29] in epithelial midgut cells. Our midgut analysis of worker honey bees demonstrates hypertrophic enlargement of the digestive cells in the first 5 days after HMF treatment. Later, ~10 days post-HMF treatment, numerous affected cells were released into the lumen, as evidenced by observable apoptotic cell death in the apical region of midgut villi. Between the 10th and 15th day of feeding, 1000 mg/kg of HMF and 1500 mg/kg of HMF increased the cell death rate, resulting in the shedding of dead cells from the epithelium into the midgut lumen. In contrast, 500 mg/kg of HMF resulted in low levels of cell death.

Initially, such changes to the apical epithelium and some cell death are adaptive, but uncontrolled epithelial cell losses without adequate regeneration leads to loss in midgut function [30,31]. In a control group where bees received only sugar candy (Apifonda), high proportions of apoptotic columnar cells were found, which indicates a high level of cell turnover. In bees exposed to high doses of HMF, normal apoptosis was followed by morphological changes typical of necrotic deletion. In this study, we found that caged bees were sensitive to the HMF treatment, as observed through changes at the tissue level, and its potential detrimental effect can result in higher bee mortality.

In addition to the differences in the mortality rates induced by HMF treatment, there was accompanying histopathology whereby lesions formed within midgut cells. HMF induced a higher cell death rate in comparison with untreated bees by the 5th, 10th, and 15th day post-treatment. Early and accelerated levels of apoptosis induced by HMF toxicity may serve as a possible defense mechanism in the midgut that impedes HMF from affecting neighboring epithelial cells, a similar mechanism to that employed when cells are infected with pathogenic organisms [32]. The midgut of honey bees also undergoes hypertrophic changes in the first day after treatment and continued to slough dead cells into the lumen. Later, a reduction in apoptotic cell deletion and cell lysis occur along the entire midgut epithelium. Thus, damaged epithelium regenerates as dying cells gradually separate from the basal lamina. The new epithelium might also form just after the complete discharging of the degenerated epithelium, when only regenerative cell groups and remnants of cell membranes are observed [33]. This regenerative process in the midgut may importantly enhance bee survival after prolonged exposure to HMF.

Clearly, HMF has a dosage-dependent cytotoxic effect on honey bee digestion; both sublethal and subclinical changes to the midgut occur at the cellular level before bees eventually die from high doses. Adverse effects of HMF feeding needs study at the colony level because bees may rely on unknown behavioral mechanisms to mitigate the toxic effects of HMF-contaminated food stores on bee and brood survival.

Author Contributions: Conceptualization, A.G. and S.J.; methodology, A.G.; investigation, S.J. and A.G.; data curation, B.S. and A.G., writing—original draft preparation, A.G. and S.J., writing—review and editing, B.S.; supervision, A.G. and B.S. Allauthors have read and agreed to the published version of the manuscript. All authors have read and agreed to the published version of the manuscript.

Funding: This research was funded by the Slovenian Research Agency, Research core funding No. P1-164; Research for improvement of safe food and health.

Acknowledgments: We are grateful for the assistance provided by Mitja Nakrst and Maja Ivana Smodiš Škerl in the experiments. The study was supported and research topics discussed by COLOSS (Prevention of honey bee COlony LOSSes, http://coloss.org/).

References

1.   White, J.W., Jr.; Doner, L. Honey Composition and Properties. In *Beesource Beekeeping*; U.S. Government Printing Office: Washington, DC, USA, 1980; pp. 82–91.

2. Hermersdörfer, H.J. Lipp: Der Honig. 3. völlig neubearbeitete und erweiterte Auflage. 205 Seiten, 66 Abbildungen, 8 Farbtafeln und 30 Tabellen. Verlag Eugen Ulmer, Stuttgart 1994. Preis: 88,-DM. *Food/Nahrung* **1995**, *39*, 540.

3. Shalumova, T.; Tanski, J.M. 5-(Hy-droxy-meth-yl)furan-2-carbaldehyde. *Acta Cryst. Sect. E Struct. Rep. Online* **2010**, *66*, o2266. [CrossRef]

4. Teixidó, E.; Santos, F.J.; Puignou, L.; Galceran, M.T. Analysis of 5-hydroxymethylfurfural in foods by gas chromatography—Mass spectrometry. *J. Chromatogr. A* **2006**, *1135*, 85–90. [CrossRef]

5. Basumallick, L.; Rohrer, J. Determination of Hydroxymethylfurfural in Honey and Biomass. *Thermo Sci. Dionex Appl. Note Update* **2001**, *10*, 6.

6. Lee, H.S.; Nagy, S. Relative reactivities of sugars in the formation of 5-hydroxymethylfurfural in sugar-catalyst model systems1. *J. Food Process. Preserv.* **1990**, *14*, 171–178. [CrossRef]

7. Gökmen, V.; Açar, Ö.Ç.; Köksel, H.; Acar, J. Effects of dough formula and baking conditions on acrylamide and hydroxymethylfurfural formation in cookies. *Food Chem.* **2007**, *104*, 1136–1142. [CrossRef]

8. Morales, F.J. Hydroxymethylfurfural (HMF) and Related Compounds. In *Process-Induced Food Toxicants: Occurrence, Formation, Mitigation, and Health Risks*; John Wiley & Sons: Hoboken, NJ, USA, 2008; ISBN 978-0-470-07475-6.

9. Tosi, E.; Ciappini, M.; Ré, E.; Lucero, H. Honey thermal treatment effects on hydroxymethylfurfural content. *Food Chem.* **2002**, *77*, 71–74. [CrossRef]

10. De Ribeiro, R.O.R.; da Carneiro, C.S.; Mársico, E.T.; Cunha, F.L.; Conte, C.A., Jr.; Mano, S.B. Influence of the time/temperature binomial on the hydroxymethylfurfural content of floral honeys subjected to heat treatment. *Ciência Agrotecnologia* **2012**, *36*, 204–209. [CrossRef]

11. Jachimowicz, T.; Sherbiny, G.E. Zur problematik der verwendung von invertzucker für die bienenfütterung. *Apidologie* **1975**, *6*, 121–143. [CrossRef]

12. Bailey, L. The Effect of Acid-Hydrolysed Sucrose on Honeybees. *J. Apic. Res.* **1966**, *5*, 127–136. [CrossRef]

13. Kammerer, F.X. Aktueller Standder Erkenntnisse über die Fütterung vonBienen mit Zucker. *Imkerfreund* **1989**, *1*, 12–14.

14. LeBlanc, B.W.; Eggleston, G.; Sammataro, D.; Cornett, C.; Dufault, R.; Deeby, T.; Cyr, E. Formation of Hydroxymethylfurfural in Domestic High-Fructose Corn Syrup and Its Toxicity to the Honey Bee (Apis mellifera). *J. Agric. Food Chem.* **2009**, *57*, 7369–7376. [CrossRef] [PubMed]

15. Van der Zee, R.; Pisa, L. Bijensterfte 2009-10 en toxische invertsuikersiroop. *NCB Rapp.* **2010**, *2*, 1–15.

16. Stabentheiner, A.; Kovac, H.; Brodschneider, R. Honeybee Colony Thermoregulation—Regulatory Mechanisms and Contribution of Individuals in Dependence on Age, Location and Thermal Stress. *PLoS ONE* **2010**, *5*, e8967. [CrossRef] [PubMed]

17. Gates, B.N. *The Temperature of the Bee Colony*; U.S. Department of Agriculture: Washington, DC, USA, 1914.

18. Evans, J.D.; Chen, Y.P.; di Prisco, G.; Pettis, J.; Williams, V. Bee cups: Single-use cages for honey bee experiments. *J. Apic. Res.* **2009**, *48*, 300–302. [CrossRef]

19. Gregorc, A.; Alburaki, M.; Werle, C.; Knight, P.R.; Adamczyk, J. Brood removal or queen caging combined with oxalic acid treatment to control varroa mites (Varroa destructor) in honey bee colonies (Apis mellifera). *Apidologie* **2017**, *48*, 821–832. [CrossRef]

20. Gregorc, A.; Sampson, B.; Knight, P.R.; Adamczyk, J. Diet quality affects honey bee (Hymenoptera: Apidae) mortality under laboratory conditions. *J. Apic. Res.* **2019**, *58*, 492–493. [CrossRef]

21. Statistical Graphics Corporation. *Statgraphics Version 5: Statistical Graphics System*; STSC Inc.: Rockville, MD, USA, 1991; ISBN 978-0-926683-06-8.

22. Bernet, D.; Schmidt, H.; Meier, W.; Burkhardt-Holm, P.; Wahli, T. Histopathology in fish: Proposal for a protocol to assess aquatic pollution. *J. Fish Dis.* **1999**, *22*, 25–34. [CrossRef]

23. Gregorc, A.; Silva-Zacarin, E.C.M.; Carvalho, S.M.; Kramberger, D.; Teixeira, E.W.; Malaspina, O. Effects of Nosema ceranae and thiametoxam in Apis mellifera: A comparative study in Africanized and Carniolan honey bees. *Chemosphere* **2016**, *147*, 328–336. [CrossRef] [PubMed]

24. Higes, M.; García-Palencia, P.; Martín-Hernández, R.; Meana, A. Experimental infection of Apis mellifera honeybees with Nosema ceranae (Microsporidia). *J. Invertebr. Pathol.* **2007**, *94*, 211–217. [CrossRef]

25. Ceksteryte, V.; Racys, J. The quality of syrups used for bee feeding before winter and their suitability for bee wintering. *J. Apic. Sci.* **2006**, *5*, 5–14.

26. Krainer, S.; Brodschneider, R.; Vollmann, J.; Crailsheim, K.; Riessberger-Gallé, U. Effect of hydroxymethylfurfural (HMF) on mortality of artificially reared honey bee larvae (Apis mellifera carnica). *Ecotoxicology* **2016**, *25*, 320–328. [CrossRef] [PubMed]

27. Okuda, K.; de Almeida, F.; Mortara, R.A.; Krieger, H.; Marinotti, O.; Tania Bijovsky, A. Cell death and regeneration in the midgut of the mosquito, Culex quinquefasciatus. *J. Insect Physiol.* **2007**, *53*, 1307–1315. [CrossRef] [PubMed]

28. Rost-Roszkowska, M.M.; Poprawa, I.; Klag, J.; Migula, P.; Mesjasz-Przybyłowicz, J.; Przybyłowicz, W. Degeneration of the midgut epithelium in Epilachna cf. nylanderi (Insecta, Coccinellidae): Apoptosis, autophagy, and necrosis. *Can. J. Zool.* **2008**, *86*, 1179–1188.

29. Prochazkova, P.; Roubalova, R.; Dvorak, J.; Tlaskalova-Hogenova, H.; Cermakova, M.; Tomasova, P.; Sediva, B.; Kuzma, M.; Bulant, J.; Bilej, M.; et al. Microbiota, Microbial Metabolites, and Barrier Function in A Patient with Anorexia Nervosa after Fecal Microbiota Transplantation. *Microorganisms* **2019**, *7*, 338. [CrossRef]

30. Chaffey, N.; Alberts, B.; Johnson, A.; Lewis, J.; Raff, M.; Roberts, K.; Walter, P. Molecular biology of the cell. 4th edn. *Ann. Bot.* **2003**, *91*, 401. [CrossRef]

31. Hill, C.A.; Pinnock, D.E. Histopathological Effects ofBacillus thuringiensison the Alimentary Canal of the Sheep Louse, Bovicola ovis. *J. Invertebr. Pathol.* **1998**, *72*, 9–20. [CrossRef]

32. James, E.R.; Green, D.R. Infection and the origins of apoptosis. *Cell Death Differ.* **2002**, *9*, 355–357. [CrossRef]

33. Rost-Roszkowska, M.M.; Machida, R.; Fukui, M. The role of cell death in the midgut epithelium in Filientomon takanawanum (Protura). *Tissue Cell* **2010**, *42*, 24–31. [CrossRef]

 *diversity*

Article

A Survey from 2015 to 2019 to Investigate the Occurrence of Pesticide Residues in Dead Honeybees and Other Matrices Related to Honeybee Mortality Incidents in Italy

Marianna Martinello, Chiara Manzinello, Alice Borin, Larisa Elena Avram, Nicoletta Dainese, Ilenia Giuliato, Albino Gallina and Franco Mutinelli *

Istituto Zooprofilattico Sperimentale delle Venezie, NRL for honeybee health, 35020 Legnaro (PD), Italy; mmartinello@izsvenezie.it (M.M.); cmanzinello@izsvenezie.it (C.M.); aliceborin@live.it (A.B.); larisaelena.avram@gmail.com (L.E.A.); ndainese@izsvenezie.it (N.D.); igiuliato@izsvenezie.it (I.G.); agallina@izsvenezie.it (A.G.)

* Correspondence: fmutinelli@izsvenezie.it; Tel.: +39-049-8084287

Received: 21 November 2019; Accepted: 24 December 2019; Published: 27 December 2019

Abstract: Honeybee health can be compromised not only by infectious and infesting diseases, but also by the acute or chronic action of certain pesticides. In recent years, there have been numerous reports of colony mortality by Italian beekeepers, but the investigations of these losses have been inconsistent, both in relation to the type of personnel involved (beekeepers, official veterinarians, members of the police force, etc.) and the procedures utilized. It was therefore deemed necessary to draw up national guidelines with the aim of standardizing sampling active ties. In this paper, we present the results of a survey carried out in Italy from 2015 to 2019, following these guidelines. Residues of 150 pesticides in 696 samples were analyzed by LC-MS/MS and GC-MS/MS. On average, 50% of the honeybee samples were positive for one or more pesticides with an average of 2 different pesticides per sample and a maximum of seven active ingredients, some of which had been banned in Europe or were not authorized in Italy. Insecticides were the most frequently detected, mainly belonging to the pyrethroid group (49%, above all tau-fluvalinate), followed by organophosphates (chlorpyrifos, 18%) and neonicotinoids (imidacloprid, 7%). This work provides further evidence of the possible relationship between complex pesticide exposure and honeybee mortality and/or depopulation of hives.

Keywords: honeybee mortality incidents; pesticide; survey; LC-MS/MS; GC-MS/MS

---

1. Introduction

In recent decades, many beekeepers from all over the world have seen a large number of their honeybee colonies dying every year [1,2]. These deaths pose a threat to global food security because honeybees, along with numerous other insect species, provide a fundamental agricultural pollination service [3,4].

Honeybees can be considered a living monitoring system of various aspects of the ecosystem. Their state of health is in fact influenced by different environmental factors, both natural and induced by human activity, such as climate trends, bee diseases, phytosanitary treatments, and beekeeping practices [5]. Honeybees commonly forage within 1.5 km of their hive (equal to an area of about 7 km^2 around the hive) and exceptionally as far as 10 or 12 km, depending on their need for food and its availability [6]. Their body is covered with hairs that can capture atmospheric residues, and they can be contaminated via food resources when gathering pollen and nectar from flowers or through water [7,8]. Consequently, during foraging flights, bees collect pollen, nectar, plant resins and water, and thus are also valid "samplers" of organic and inorganic chemicals in the environment, which are often taken back to the colony.

In Italy, since 2003, significant honeybee mortality has been recorded in springtime, mainly related to the side-effects of maize seed dressed with neonicotinoid insecticides [9]. These events increased significantly in 2008 leading to the creation, in 2009, of a nationwide surveillance network to monitor the health status of beehives and to properly report bee death incidents and their possible causes. The monitoring network, named ApeNet (2009–2010), initially included about 100 apiaries distributed throughout most of Italy, increasing to 300 and a total of approximately 3000 beehives in 2011 with the BeeNet project (2011–2014), both funded by the Italian ministry of Agriculture, Food and Forestry Policies [10]. These projects have made it possible to assess the health status of hives in Italy, through field observations, surveys, and laboratory analyses aimed at identifying specific pathogens and chemicals, and to study episodes of honeybee colony mortality. A bee emergency service team (BEST) has been created, in charge of receiving beekeepers' reports, assessing severity, organizing and participating in investigations, or coordinating the technicians recruited to deploy them in agreement with the competent authorities [11]. During the five-year monitoring studies, annual and regional variations were observed in pathogens responsible for infection (Deformed Wing Virus, Acute Bee Paralysis Virus, Chronic Bee Paralysis Virus, *Nosema ceranae*), and in *Varroa* mite prevalence [2]. Bee bread was often contaminated with at least one pesticide and the number of detected pesticides was positively related to the size of the agricultural area surrounding the apiaries [12]. Of the honeybee samples received following the application of the BEST protocols, 126 were analyzed, of which approximately 50% were positive for at least one active ingredient. The most frequently detected pesticides were imidacloprid, chlorpyrifos, thiachloprid, clothianidin, and thiametoxam.

In 2014, the Italian ministry of Health implemented the following regulations: (1) Regulation (EC) No. 1107/2009 [13] concerning the placing of plant protection products (PPPs) on the market (repealing Council Directives 79/117/EEC and 91/414/EEC); (2) Directive 2009/128/EC [14] establishing a framework for Community action to achieve the sustainable use of pesticides; (3) Commission Directive 2010/21/EU amending Annex I to Council Directive 91/414/EEC regarding specific provisions relating to clothianidin, thiamethoxam, fipronil and imidacloprid [15]. Commission Directive 2010/21/EU indicates that member states shall ensure monitoring programs are initiated where and as appropriate to verify the real exposure of honeybees to the aforementioned neonicotinoids in areas extensively used by bees for foraging or by beekeepers. Furthermore, in view of the still frequent beekeeper reports of honeybee death at certain times of the year, and the high degree of public attention paid to this problem, it was considered appropriate for the various Italian regions to adopt a more systematic approach to the management (notification to the competent authority, epidemiological and clinical investigation in the apiary, sampling, laboratory investigation) of bee mortality incidents where pesticide poisoning was suspected. In fact, the application of heterogeneous procedures in the management of bee killing incidents could result in data that may not be representative or exhaustive. Consequently, the General Directorate of Animal Health and Veterinary Medicinal Products of the Italian ministry of Health, with note number 0016168 dated 31 July 2014 [16] issued the "Linee guida per la gestione delle segnalazioni di moria o spopolamento degli alveari connesse all'utilizzo di agrofarmaci" (Guidelines for the management of reports of death or depopulation of bee colonies related to the use of plant protection products). These guidelines provide operational directives for managing these events with the aim of helping to protect beekeeping heritage from poisoning by plant protection products (PPPs), gathering information on the possible causes of death and/or depopulation of beehives, standardizing investigations in terms of the procedures adopted both in the field and at the laboratories responsible for analyzing the sampled dead bees.

The aim of the present study was to investigate the presence of pesticide residues in dead honeybees submitted to our laboratory following the guidelines for managing reports of death or depopulation of bee colonies related to the use of PPPs in Italy, from 2015 to 2019. Sample extraction was based on the QuECheRS technique followed by liquid or gas chromatography, both coupled with mass spectrometry (LC-MS/MS and GC-MS/MS), to analyze the selected active substances.

2. Materials and Methods

2.1. Sampling

Dead honeybee sampling and shipment were carried out by official veterinarians of the Local Health Service specifically trained in beekeeping, in the presence of the beekeeper who issued the notification. The official veterinarians operated in accordance with the abovementioned guidelines, defining the working protocol with all monitoring details, to further standardize the procedure across the different apiaries and beekeepers. Samples consisted of 250–1000 dead honeybees, collected in suitably sealed, properly identified containers. Optionally, samples of comb, bee bread, and vegetable matrices (most frequently leaves, corn seedlings, maize) were also collected. The samples were individually packed in plastic sampling bags to avoid cross contamination, properly identified and the proper storage was guaranteed by immediate freezing after collection. Dead honeybees were collected at the hive entrance or from the ground in front of the hive, bee bread was taken directly from the comb as well as honey. All beehives were opened and clinically inspected in order to evaluate the size of colonies and to estimate the possible impact of the bee killing incidents on the colony itself. Vegetable matrices were collected in the immediate vicinity of the hive or from the near crops treated with pesticides. All samples were stored at −20 °C until delivery to the territorially competent Veterinary Institute (IZS) and until toxicological analysis. Samples considered in the present work were delivered to our laboratory between 2015 and 2019. Figure 1 depicts the location of the sampling points in Italy for all matrices and mortality events by year.

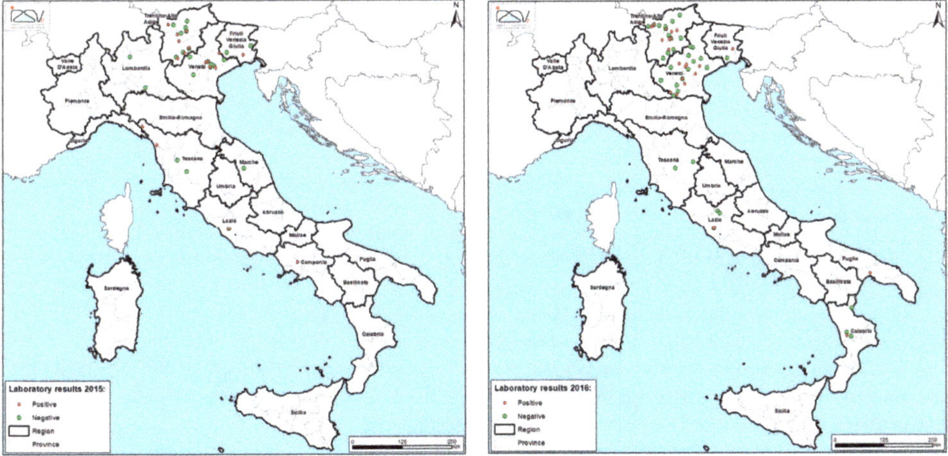

Figure 1. *Cont.*

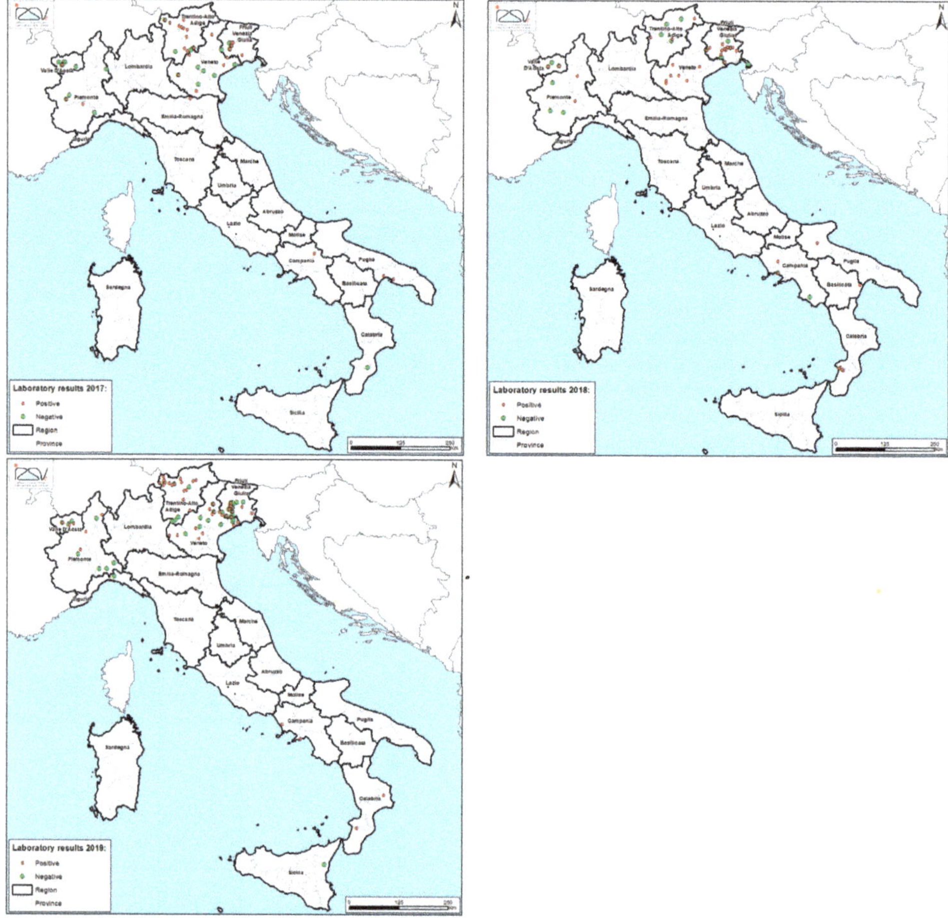

Figure 1. Location of positive and negative samples for each officially reported honeybee mortality event in Italy from 2015 to 2019.

2.2. Chemicals

Analytical-grade (98–99.9% purity) standards of pesticides were supplied by Sigma-Aldrich (Steinheim, Germany) and are listed in Appendix A (Table A1). Pesticide-grade solvents, Supel™ QuE Citrate Extraction Tubes and Supel™ QuE PSA/C18 (EN) Cleanup Tubes were used to extract and purify samples and were purchased from Sigma-Aldrich (Steinheim, Germany). High purity water was prepared using a Milli-Q water purification system (Millipore, Milford, MA, USA). Single standard solutions were prepared in methanol at a concentration of 1000 mg/L. The working standard solutions were prepared by mixing the appropriate amounts of single standard solutions and diluting with methanol to a final concentration of 10 and 1 mg/L. All solutions were stored in the dark in 10 mL amber bottles at −20 °C.

2.3. Sample Preparation

Samples were prepared following the QuEChERS (Quick Easy Cheap Effective Rugged Safe) approach using the slightly modified method reported by Anastassiades et al. [17]. To obtain proper homogenization and extraction, the samples were previously pulverized with a crushing mill

(A11 basic IKA-Werke GmbH & Co. KG, Staufen, Germany) cooled with liquid nitrogen. The samples were processed in duplicate as they were subsequently analyzed with both LC and GC techniques. For extraction, one gram of pollen and vegetable matrices, two grams of bees and wax or 5 g of honey, were weighed into a centrifuge tube and 10 mL of water was added. The mixture was vortexed for 5 min, acetonitrile with 0.1% acetic acid (10 mL) was added, vortexed for 20 min and cooled at −20 °C for 15 min. To perform the partitioning step, QuEChERS salts EN method (sodium citrate 1 g, sodium hydrogen citrate sesquihydrate 0.5 g, magnesium sulphate 4 g and sodium chloride 1 g) were added and vigorously shaken up and down for 1 min. The mixture was centrifuged and 7 mL of supernatant was transferred to a tube containing purification dispersive SPE Fatty Samples EN salts (magnesium sulphate 900 mg, PSA 150 mg and C18 150 mg). The solution was vortexed for 1 min and centrifuged, and 4 mL of the supernatant was transferred to a clean tube and evaporated to dryness under vacuum at 45 °C. The residue was dissolved in 0.5 mL of reconstitution solution, composed of 5 mM ammonium formiate in water with 0.1% formic acid and 5 mM ammonium formiate in methanol with 0.1% formic acid (1:1 *v/v*), and PTFE filtered (0.45 μm pore size) for analysis by UPLC-MS/MS (Ultra Pressure Liquid Chromatography coupled with tandem mass spectrometry). Samples analyzed using GC-MS/MS were reconstituted with 0.5 mL of heptane and PTFE filtered (0.45 μm pore size). Both instruments were programmed in MRM (multiple reaction monitor) mode with two selected transitions per molecule.

2.4. LC-MS/MS Analysis

The analysis was performed using a Shimadzu LCMS-8040 (Kyoto, Japan), with a tandem quadrupole analyzer, in MRM spectrum mode using an electron spray ionization source in both positive and negative ionization modes. The chromatography was performed on a Raptor (Restek Corporation, Bellefonte, PA, USA) biphenyl column (10 cm × 2.1 mm, 2.7 μm-particles) with an adequate guard column, thermostated at 35 °C. The mobile phase solvents were 5 mM ammonium formiate in water with 0.1% formic acid and 5 mM ammonium formiate in methanol with 0.1% formic acid. The chromatographic eluting conditions were optimized as follows: from 3% to 10% B (0–1 min), from 10% to 55% B (1–3 min), from 55% to 100% B (3–10.5 min), 100% B maintained for 2.5 min, from 100% to 3% B in 0.01 min), followed by re-equilibration to 3% B for a further 3 min. The total analysis run time was 15 min. The flow rate was 0.4 mL/min and the injection volume was 2 μL. Quantitative and qualitative analyses were performed with LabSolution Insight software based on the two most intensive fragment ion transitions. The matrix matched standards were used for calibration and quantification, prepared by analyzing blank (negative) samples spiked with pesticides after the extraction and purification steps.

2.5. GC-MS/MS Analysis

The analysis was performed on a Shimadzu GC-MS TQ8040 equipped with Phenomenex ZB-Semivolatiles columns (30 m, 0.25 mm ID, 0.25 μm) and a tandem mass spectrometry detector. A sample volume of 1 μl was injected in the splitless mode at an injector temperature of 270 °C. The oven temperature was programmed as follows: initial temperature 60 °C (held for 2 min) increased by 70 °C/min to 200 °C; increased by 6 °C/min to 300 °C (held for 2 min). The ion source and interface temperature were held at 230 °C and 280 °C, respectively. The total analysis run time was 23 min. Quantitative and qualitative analyses were performed with LabSolution Insight software based on the two most intensive fragment ion transitions. The matrix matched standards were used for calibration and quantification, prepared by analyzing blank samples spiked with pesticides after the extraction and purification steps.

3. Results

Table 1 presents a summary of the pesticides detected in the samples analyzed within 5 years of monitoring. Tables 2 and 3 summarize the main findings by survey year for honeybees and other matrices, respectively. Figure 1 shows the location of the positive and negative samples for each officially reported honeybee mortality event.

Table 1. Active ingredients detected in the samples analyzed in the five years of monitoring.

Active Ingredient	Pesticide Type [a]	Substance Group	Honeybees		Other Matrices [b]	
			Prevalence (%)	Range (ng per Bee)	Prevalence (%)	Range (mg kg^{-1})
3-ketocarbofuran	metabolite	carbamate	-	-	0.7	0.04
acetamiprid	I	neonicotinoid	0.8	1.9–6.6	1.7	0.03–0.06
acrinathrin	I-A	pyrethroid	2.0	42.7–473	1.7	0.2–0.8
azoxystrobin	F	strobilurin	1.2	1.0–5.4	2.5	0.01–0.02
boscalid	F	carboxamide	1.2	<LOQ -25.3	0.8	0.02
bromopropylate	A	benzilate	1.2	5.7–14.5	-	-
bupirimate	F	pyrimidinol	0.8	1.6–1.8	-	-
carbendazim	F	benzimidazole	0.4	12.9	0.8	0.04
chlormequat chloride	PGR	quarternary ammonium compound	0.4	1.9	0.8	0.02
chlorothalonil	F	chloronitrile	0.4	84.8	-	-
chlorfenvinphos	I-A	organophosphate	2.8	<LOQ -162.0	7.6	0.01–0.4
chlorpyrifos	I	organophosphate	12.9	1.0–1688.8	8.5	0.01–0.5
chlorpyrifos-mehyl	I-A	organophosphate	6.8	1.1–316.0	1.7	0.01
chlorpropham	I-PGR	carbamate	0.8	7.2	-	-
clomazone	H	isoxazolidinone	0.4	1.1	-	-
clothianidin	I	neonicotinoid	3.6	1.0–17.0	-	-
coumaphos	I	organothiophosphate	1.6	1.5–50.0	10.2	0.01–0.7
cymoxanil	F	cyanoacetamide oxime	-	-	0.8	0.03
cypermethrin	I	pyrethroid	2.8	1.1–7602.0	-	-
cyprodinil	F	anilinopyrimidine	2.8	<LOQ -484.5	0.8	1.1
deltamethrin	I-M	pyrethroid	-	-	3.4	0.05–0.1
desmethyl-pirimicarb	I	carbamate	0.4	1.7	2.5	0.03–0.1
dimethoate	I-A	organophosphate	3.2	23.3–1647.4	1.7	
dimethomorph	F	morpholine	3.2	<LOQ -4.3	9.3	0.02–47.0
dodine	F	guanidine	-	-	1.7	0.01–0.2
ethoprophos	I-N	organophosphate	0.4	1.7	1.7	0.2
etofenprox	I	pyrethroid	6.0	<LOQ -17.5	2.5	0.01–0.03
fipronil	I	phenylpyrazole	0.4	139.6	0.8	1.8
fludioxonil	F	phenylpyrrole	1.2	1.0–6.5	0.8	0.01
fluopicolide	F	benzamide	0.8	3.1	-	-
fluopyram	F	benzamide, pyramide	2.0	4.0–6.7	-	-
flutriafol	F	triazol	0.4	2.0	-	-
folpet	F	phthalimide	2.4	9.1–46.0	4.2	0.3–511.1
imidacloprid	I	neonicotinoid	11.6	1.2–402.6	3.4	0.02–4.1
indoxacarb	I	oxadiazine	0.8	2	-	-
iprodione	F	dicarboximide	1.2	3.0–96.3	0.8	0.03
kresoxim-methyl	F-B	strobilurin	0.8	2.0–4.2	0.8	0.01
metalaxil	F	phenylamide	2.0	1.0–9.0	5.1	0.04–0.4
metalaxyl-M	F	phenylamide	-	-	0.8	0.03
methiocarb	I	carbamate	4.8	<LOQ -46.5	11.0	0.01–70.6
methiocarb sulfoxide	metabolite	carbamate	2.4	2.5–5.4	2.5	0.01–7.6
methomyl	I-A	carbamate	3.6	1.0–765.0	2.5	0.2–0.3
metrafenone	F	benzophenone	0.8	1.4–2.8	0.8	0.03
metribuzin	H	triazinone	-	-	0.8	0.08
omethoate	I-A	organophosphate	3.2	4.7–102.2	-	-
penconazole	F	triazole	6.4	1.0–90.0	2.5	0.01–0.02
pendimethalin	H	dinitroaniline	2.0	1.2–4.3	2.5	0.01–0.03
permethrin	I	pyrethroid	13.3	1.6–134,665.0	5.1	0.02–2.0
phosmet	I-A	organophosphate	3.6	1.0–280.8	1.7	0.2–0.7
piperonil butoxide	-	cyclic aromatic	11.2	1.0–66,827.0	11	0.01–2.3
propamocarb	F	carbamate	2.8	1.1–8.5	2.5	0.02–3.6

Table 1. *Cont.*

Active Ingredient	Pesticide Type [a]	Substance Group	Honeybees		Other Matrices [b]	
			Prevalence (%)	Range (ng per Bee)	Prevalence (%)	Range (mg kg^{-1})
propiconazole	F	triazole	2.4	1.4–6.0	-	-
pyraclostrobin	F	strobilurin	0.4	6.7	1.7	0.01–0.3
pyrimethanil	F	anilinopyrimidine	4.4	1.0–89.0	6.8	0.02–0.8
pyriproxyfen	I	unclassified	0.8	3.7–5.9	-	-
quinoxyfen	F	quinoline	1.6	2.8–13.5	1.7	0.01
rotenone	I-A	isoflavone	0.8	1.5–2.4	0.8	0.04
s-metolachlor	H	chloroacetamide	-	-	0.8	0.6
tau-fluvalinate	I-A	synthetic pyrethroid	38.2	1.0–1018.0	53.4	0.02–95.9
tefluthrin	I	pyrethroid	3.6	1.1–33.1	4.2	0.07–1.6
terbuthylazine	H	triazine	-	-	4.2	0.02–1.0
tetraconazole	F	triazole	-	-	4.2	0.01–1.1
tetramethrin	I	pyrethroid	8.4	1.0–71,096.0	2.5	0.03–0.9
thiacloprid	I	neonicotinoid	4.0	1.1–9.5	1.7	0.01–0.04
thiamethoxam	I	neonicotinoid	1.6	1.8–33.6	-	-
thifensulfuron-methyl	H	sulfonylurea	0.4	2.1	-	-
thiodicarb	I	carbamate	0.8	2.0–5.0	-	-
thiophanate-methyl	F	benzimidazole	0.8	5.8–78.6	0.8	0.2
tribenuron-methyl	H	sulfonylurea	0.4	1.0	-	-
trifloxystrobin	F	strobilurin	0.8	8.1–19.9	0.8	1.4

[a] A, acaricide; B, bactericide; F, fungicide; H, herbicide; I, insecticide; N, nematicide PGR, plant growth regulator. [b] beeswax, beebread, soil, leaves and maize seeds.

Table 2. Pesticides detected in the samples of honeybees analyzed in the five years of monitoring.

Year	n. of Samples	n. of Positive Samples (%)	n. of Pesticides Detected	Most Frequent	Prevalence (%)
2015	68	28 (41)	20	fluvalinate	50.0
				piperonyl butoxide	25.0
				cypermethrin	14.3
				chlorpyrifos	14.3
2016	112	47 (42)	22	chlorpyrifos	25.5
				permethrin	25.5
				tetramethrin	25.5
				piperonil butoxide	23.4
				fluvalinate	21.3
2017	95	48 (50)	22	fluvalinate	41.7
				chlorpyrifos	37.5
				methomyl	16.7
2018	85	63 (74)	36	fluvalinate	34.9
				imidacloprid	14.3
				etofenprox	12.7
				methiocarb	12.7
2019	147	63 (43)	34	fluvalinate	46.0
				chlorpyrifos	15.9
				imidacloprid	14.3
Sum	507	249 (50)	63	fluvalinate	38.2
				permethrin	13.3
				chlorpyrifos	12.9

Table 3. Pesticides detected in other matrices analyzed in the five years of monitoring.

Year	n. of Samples	Matrix	n. of Positive Samples (%)	n. of Pesticides Detected	Most Frequent	Prevalence (%)
2015	18		12 (67)	12	tau-fluvalinate	83.3
					tetraconazole	25.0
	15	comb	10 (67)	10	tau-fluvalinate	100.0
	1	honey	1 (100)	2		
	2	vegetable	1 (50)	3		
2016	39		22 (56)	15	tau-fluvalinate	45.5
					metalaxyl	13.6
					permethrin	13.6
	28	comb	15 (54)	7	tau-fluvalinate	67
	3	pollen	2 (67)	1	metalaxyl	100
	3	honey	0			
	5	vegetable	3 (60)	6		
2017	23		13 (57)	10	tau-fluvalinate	30.8
					chlorpyrifos	23.1
					desmethyl-pirimicarb	15.4
	11	comb	10 (91)	9	tau-fluvalinate	60.0
	7	pollen	3 (43)	3	tau-fluvalinate	66.7
	5	honey	0			
2018	36		30 (83)	26	tau-fluvalinate	50
					methiocarb	50
					coumaphos	26.7
	11	comb	10 (91)	10	tau-fluvalinate	100.0
	10	pollen	9 (90)	14	tau-fluvalinate	66.7
	7	honey	5 (71)	1	methiocarb	100.0
	8	vegetable	6 (75)	12	deltamethrin	66.7
2019	73		41(56)	27	tau-fluvalinate	58.5
					pyrimethanil	19.5
					dimetomorph	17.1
	32	comb	25 (78)	16	tau-fluvalinate	68.0
	14	pollen	11 (78)	11	tau-fluvalinate-pyrimethanil	27.3
	25	honey	3 (12)	2	tau-fluvalinate	66.7
	2	vegetable	2 (100)	6		
Sum	189		118 (67)	51	tau-fluvalinate	53.4
					methiocarb	13.6
					piperonyl butoxide	11.0

In total, 63 different active ingredients were found in honeybee samples, with concentrations ranging from 0.1 to 134,665 ng/bee, and 51 different active ingredients in the other analyzed matrices (beeswax, bee bread, honey and vegetable matrices), ranging from 0.01 to 359.5 mg/kg. Most investigated samples were positive for at least one active ingredient (53%) and contaminated by more than one residue: 53% of the samples were contaminated by at least two different residues, 32% by at least three, while as many as nine active ingredients were detected in one extreme case, coming from Udine province (North-Eastern Italy).

Insecticides were the most frequently detected active substances (49.2%) in honeybees (Tables 1 and 2), the most prevalent being the acaricide tau-fluvalinate (38.2%). Pyrethroid permethrin, the second most frequently found active substance, had a prevalence of 13.3%. Chlorpyrifos was the third most commonly determined pesticide (12.9%).

Globally, most of the other matrices analyzed (67%) were positive for at least one active ingredient (Table 3). Again, the acaricide tau-fluvalinate was the most commonly found active ingredient, with a prevalence of 53.4%, followed by the insecticide methiocarb with a prevalence of 13.6% (also considering the metabolite methiocarb sulfoxide), and the synergist piperonyl butoxide (11.0%).

4. Discussion

The geographical distribution of the honeybee death events (Figure 1) coincides with the areas in Italy in which intensive agriculture is mainly practiced (such as apple and citrus orchards and vineyards mainly other than maize cultivations).

In honeybees (Tables 1 and 2), the most frequently detected active substances were insecticides with tau-fluvalinate having the highest prevalence. Tau-fluvalinate is a pyrethroid insecticide authorized both as a PPP and for the control of *Varroa* mite infestation of honeybees in Italy. Miticides have already been found by different studies [2,18,19] to be the most frequent residues in honeybee samples around Europe. The pyrethroid permethrin, the second most frequently identified active ingredient, also has the highest detected concentration. It is a contact insecticide which has not been approved for use in the EU as a PPP, due particularly to its acute toxicity to aquatic organisms. Chlorpyrifos was the third most commonly determined pesticide in honeybees and, being an active ingredient highly toxic to bees, it could represent an important factor affecting colony health. Chlorpyrifos has already been identified as one of the most commonly detected insecticides in bees [20–22]. Neonicotinoids, mainly imidacloprid, were also frequently identified. In Italy, the use of three neonicotinoids, namely imidacloprid, clothianidin, thiamethoxam, and fipronil, was restricted in 2008 due to evidence of their negative effects on honeybee health. In 2013 the EU definitively banned the use of these active ingredients for seed treatment, soil application and foliar treatment of plants and cereals attractive to bees (but use in greenhouses is allowed) [23,24]. However, fipronil and all three neonicotinoids (including the restricted ones) screened in our study were detected in our samples. Therefore, despite the current limitations on the use of PPPs containing these active ingredients, according to the present monitoring results, honeybees are still exposed to potentially harmful levels of these pesticides, as already observed in previous studies [25–27]. Fungicides were also often detected (39.3%) with a wide variety of active ingredients, the most frequently found being penconazole and pyrimethanil. Although there are no restrictions on the use of fungicides on crops during blooming, various studies have shown that the impact of fungicides on honeybee health can be harmful, both due to their direct negative effects on honeybee health [28,29], and through a synergistic action between fungicides and other types of pesticide [30–32]. Our results partially agree with those obtained in a previous study carried out by our laboratory [33], which assessed the presence of pesticides and viruses in dead honeybees following mortality incidents in northeastern Italy in 2014. Compared to this study, in which imidacloprid was the most frequently detected active ingredient, there has now been a reduction in the presence of neonicotinoids, probably due to limitations imposed on their use by the European Commission [23,24]. Tau-fluvalinate and chlorpyrifos were instead confirmed to be among the most frequently identified active ingredients.

With the exception of 2018, the year in which we observed close correspondence between honeybees positive to pesticides and bee kill incident reports, percentage positivity stood at around 44% in the other monitoring years: a value probably influenced by various factors, as the speed of reporting and the subsequent sampling intervention. The concentration of pesticides in dead honeybees can rapidly decrease within just hours of the poisoning event and, if not properly stored at −20 °C, samples can reach a level close to environmental residue before being analyzed in the laboratory [20]. The analysis results may be also affected by the severity of the poisoning event (in terms of the active ingredients involved, their concentration, method of administration) and the presence of other bee parasites or stressors (such as viruses and *Varroa* mite) that can contribute to the weakening of colonies and predisposition to mortality events, even with sublethal concentrations of pesticides [28,34,35].

The honeybee is certainly the most important matrix to be analyzed in case of honeybee mortality incidents, as residues detected in honeybees reflect their exposure both to direct contact with PPPs, biocides, or even veterinary drugs, and to the consumption of contaminated nectar and pollen. However, the analysis of other matrices related to the same incident can help us to better understand the mortality event. For example, bee bread can supply useful data on any PPPs application occurring in the areas surrounding the beehive, while beeswax comb can provide information on exposure over a period of time. Unlike other beehive products, beeswax can remain in the hive for many years, thus resulting in an accumulation of various non-polar active ingredients applied in beekeeping and agriculture [19,35]. In the present study, most other analyzed matrices (67%) were positive for at least one active ingredient (Table 3), and again the acaricide tau-fluvalinate represented the most commonly found active substance, followed by the insecticide methiocarb and the synergist piperonyl butoxide. The most represented matrix was beeswax, with an average of 72% (70 out of a total of 97 samples) of the samples proving to be contaminated with pesticides, mainly tau-fluvalinate. Bee bread showed 74% positivity (25/34), and in this case too, the most commonly detected active ingredient was tau-fluvalinate. Being stored inside the beehive, bee bread can be affected by both beekeeper and agricultural activity. For these reasons, however, in the case of a honeybee killing event, we cannot rely on toxicological information provided by beeswax. Vegetable matrices (most frequently leaves, corn seedlings, maize) were contaminated in 70% of cases (12/17), with the widest variety of active ingredients (27), despite being by far the least numerous matrix received. Honey was also received as a matrix related to honeybee incidents and proved to be contaminated with pesticides in only 20% of cases; but in three samples the detected pesticide concentration exceeded the limits imposed by the EU (methiocarb 0.05 and 0.7 mg/kg and tau-fluvalinate 0.05 mg/kg) [36]. These results should draw attention to the fact that mortality events are harmful to honeybees, but consumers' health should also be considered. The risk of contamination of edible beehive products, as honey and pollen, but also beeswax, which can then be reused and lead to the transfer of contaminants to honey, cannot be ruled out [37].

It is also worth mentioning the detection of some active ingredients that are no longer authorized but in the past were present in both PPPs and veterinary medicinal products. Authorized active ingredients used against varroosis [38] were among the main sources of honeybee and hive matrices contamination, but so were old apicultural and agricultural acaricides that are now banned, such as bromopropylate (both), chlorfenvinphos, and rotenone (agricultural). The pyrethroid insecticide permethrin, which is highly toxic to honeybees and authorized as a biocide [39], was frequently detected, even in high concentrations, in both honeybees and other matrices. The same applies to the potent multi-purpose pyrethroid insecticide tetramethrin, registered in 1968 and often used to control insects presenting risks to public health, but which is highly toxic for honeybees and has never been authorized for use in crop protection. The insecticide thiodicarb was detected in a few honeybee samples. This insecticide and molluscicide is used to control Lepidoptera, Coleoptera, slugs, other pests of fruit, vegetables, and many other crops, with moderate or high toxicity to honeybees, depending on whether the administration is contact or oral [36,40].

5. Conclusions

The data collected following the five-year monitoring survey showed that the application of ministerial guidelines allows the gathering of data on honeybee mortality incidents at national level in a consistent and reliable manner. We have shown that honeybee mortality events are still occurring and widespread, and that honeybees and beehive products are widely exposed to a large number of substances used legally and illegally, in agricultural practices and in beekeeping. In the honeybee matrix, 50% of the samples were found to be positive, while a greater proportion of the other matrices were contaminated. The honeybee is certainly the most interesting matrix for this study but also the most delicate from an analytical point of view, considering that laboratory results may be affected by various factors, from meteorological aspects to beekeeper reporting times and consequently the intervention of the official veterinarian responsible for sampling. This could potentially result in an underestimation of the problem. As a consequence, beekeepers and official veterinarians need to be highly aware and well informed of this problem to ensure that reporting and samplings are as punctual and prompt as possible. It is also important for the laboratory assigned to sample analysis to be aware of the problems linked to the possibility of pesticide concentration decreasing rapidly in dead bees and therefore of the best ways to conserve the samples before analysis. The pesticide panel must also be kept up to date, based on the continuous evolution of the pesticides available on the market. Furthermore, the notification of honeybee killing incidents to the competent national and regional authorities could contribute to increase the awareness of farmers about the possible impact on honeybees of PPPs application. Moreover, this awareness could lead to a more reasonable application of the mitigation measures (established at regional level), such as proper maintenance of PPP application machines together with the use of deflectors to reduce the drifting of active ingredients during treatment, as well as to cut the grass on the orchard or vineyard surface when blossoms are present. The latter measure could strongly reduce the risk of exposure of honeybees to contaminated sources of nectar and pollen, even when the orchard is not blooming.

Our results, based on the appropriate management of bee killing events, as described above, together with laboratory investigations, could contribute to a better understanding of the influence of pesticide mixtures on honeybee health, even at sublethal concentrations. The application, for example, of otherwise sublethal doses of miticides when tau-fluvalinate and coumaphos are simultaneously present in the hive could lead to honeybee mortality [41]. Likewise, great synergy is observed in the laboratory between EBI fungicides at field application rates and pyrethroids used as varroacides [42]. The present type of forensic study cannot demonstrate a direct link between honeybee mortality and pesticide mixtures but does provide us with valid indications of the interactions between active ingredients and therefore the pesticides that warrant further study in the future.

Author Contributions: Conceptualization, F.M., M.M.; Methodology, M.M., A.G.; Validation, M.M. and A.G.; Investigation, M.M., C.M., A.B., L.E.A., N.D., I.G.; Data Curation, C.M.; Writing—Original Draft Preparation, M.M.; Writing—Review & Editing, F.M., M.M.; Supervision, F.M. All authors have read and agreed to the published version of the manuscript.

Funding: This research received no external funding.

Acknowledgments: The authors wish to thank Claudia Casarotto, GIS Office of Istituto Zooprofilattico Sperimentale delle Venezie, for preparing the maps and Andrea Maroni Ponti, Italian ministry of Health, Directorate General of animal health and veterinary medicines for supporting the program.

Conflicts of Interest: The authors declare no conflict of interest.

Appendix A

Table A1. Summary of the active substances studied (LOQ = 10 µg kg^{-1}).

Active Substances	Substance Group	Pesticide Type [a]	Detection Method
Acetamiprid	neonicotinoid	IN	HPLC-MS
Acrinathrin	pyrethroid	IN-AC	GC-ECD
Aldicarb *	carbamate	IN-AC-NE	HPLC-MS
Aldicarb-sulfone (Aldicarb metabolite) *	carbamate	IN-AC-NE	HPLC-MS
Aldicarb-sulfoxide (Aldicarb metabolite) *	carbamate	IN-AC-NE	HPLC-MS
Alpha-Endosulfan *	organochlorine	IN-AC	GC-ECD
Azoxystrobin	strobilurin	FU	HPLC-MS
Benalaxyl	Acylalanine	FU	HPLC-MS
Beta-Endosulfan *	organochlorine	IN-AC	GC-ECD
Bifenthrin	pyrethroid	IN-AC	GC-ECD
Bitertanol *	triazole	FU	HPLC-MS
Boscalid	carboxamide	FU	HPLC-MS
Bromopropylate *	benzilate	AC	HPLC-MS
Bupirimate	pyrimidinol	FU	HPLC-MS
Captan	phtalimide	FU-BA	GC-ECD
Carbaryl *	carbamate	IN-PG	HPLC-MS
Carbendazim	benzimidazole	FU	HPLC-MS
Carbofuran *	carbamate	IN-AC-NE	HPLC-MS

Table A1. *Cont.*

Active Substances	Substance Group	Pesticide Type [a]	Detection Method
Carbofuran-3-hydroxy (Carbofuran metabolite) *	carbamate	IN-AC-NE	HPLC-MS
Carbofuran-3-keto (Carbofuran metabolite) *	carbamate	IN-AC-NE	HPLC-MS
Carboxine	oxathiin	FU	HPLC-MS
Chlorfenvinphos *	organophosphate	IN-AC	GC-ECD
Chloridazon	pyridazinone	HB	HPLC-MS
Chlormequat chloride	quaternary ammonium	PG	HPLC-MS
Chlorothalonil	chloronitrile	FU	GC-ECD
Chlorpropham	carbamate	HB-PG	GC-ECD
Chlorpyrifos	organophosphate	IN	GC-ECD
Chlorpyrifos-methyl	organophosphate	IN-AC	HPLC-MS
Clomazone	isoxazolidinone	HB	HPLC-MS
Clothianidin	neonicotinoid	IN	HPLC-MS
Coumaphos *	organophosphate	IN-AC	GC-ECD
Cyfluthrin-beta	pyretroid	IN	GC-ECD
Cyhalothrin *	pyretroid	IN	GC-ECD
Cypermethrin	pyretroid	IN	GC-ECD
Cyproconazole	triazole	FU	HPLC-MS
Cyprodinil	anilinopyrimidine	FU	HPLC-MS
Deltamethrin	pyretroid	IN	GC-ECD
Diflubenzuron	benzoylurea	IN	HPLC-MS
Dimethoate	organophosphate	IN-AC	HPLC-MS
Dimethomorph	morpholine	FU	HPLC-MS
Dinotefuran	neonicotinoid	IN	HPLC-MS
Dithianon	quinone	FU	HPLC-MS
Dodemorph	morpholine (isomer mix)	FU	HPLC-MS
Dodine	guanidine	FU	HPLC-MS
Endosulfan sulfate	organochlorine	IN-AC	GC-ECD
Esfenvalerate	pyretroid	IN	GC-ECD
Ethoprophos	organophosphate	IN-NE	GC-ECD
Etofenprox	pyretroid	IN	HPLC-MS
Etoxazole	diphenyl oxazoline	AC	HPLC-MS
Etridiazole	aromatic hydrocarbon	FU	GC-ECD
Fenamidone	imidazole	FU	HPLC-MS
Fenamiphos	organophosphate	NE	HPLC-MS
Fenamiphos sulfone (Fenamiphos metabolite)	organophosphate	NE	HPLC-MS
Fenamiphos sulfoxide (Fenamiphos metabolite)	organophosphate	NE	HPLC-MS
Fenarimol *	pyrimidine	FU	HPLC-MS
Fenazaquin	quinazoline	IN-AC	HPLC-MS
Fenbuconazole	triazole	FU	HPLC-MS
Fenothiocarb *	thiocarbamate	AC	HPLC-MS
Fenoxycarb	carbamate	IN	HPLC-MS
Fenpropidin	piperidine	FU	HPLC-MS
Fenpropimorph	morpholine	FU	HPLC-MS
Fenpyroximate	pyrazole	IN-AC	HPLC-MS
Fenvalerate *	pyretroid	IN-AC	GC-ECD
Fipronil	phenylpyrazole	IN	GC-ECD
Flazasulfuron	sulfonylurea	HB	HPLC-MS
Fluazifop-P-butyl	aryloxyphenoxypropionate	HB	HPLC-MS
Flufenacet	oxyacetamide	HB	HPLC-MS
Flufenoxuron *	benzoylurea	IN-AC	HPLC-MS
Fludioxonil	phenylpyrrole	FU	HPLC-MS
Fluopicolide	benzamide	FU	GC-ECD

Table A1. *Cont.*

Active Substances	Substance Group	Pesticide Type [a]	Detection Method
Fluopyram	benzamide, pyramide	FU-NE	GC-ECD
Fluquinconazole	triazole	FU	GC-ECD
Flusilazole	triazole	FU	HPLC-MS
Flutriafol	triazole	FU	HPLC-MS
Folpet	phtalimide	FU	GC-ECD
Forchlorfenuron	phenylurea	PG	HPLC-MS
Formetanate	formamidine	IN-AC	HPLC-MS
Fosthiazate	organophosphate	IN-NE	HPLC-MS
Imazalil	imidazole	FU	HPLC-MS
Imazosulfuron	sulfonylurea	HB	HPLC-MS
Imidacloprid	neonicotinoid	IN	HPLC-MS
Indoxacarb	oxadiazine	IN	GC-ECD
Iprodione	dicarboximide	FU	GC-ECD
Iprovalicarb	carbamate	FU	HPLC-MS
Isopyrazam	pyrazole	FU	GC-ECD
Kresoxim-methyl	strobilurin	FU	GC-ECD
Lambda-cyhalothrin	pyrethroid	IN	GC-ECD
Linuron	urea	HB	HPLC-MS
Malaoxon (Malathion metabolite) °	organophosphate	IN-AC	HPLC-MS
Malathion	organophosphate	IN-AC	HPLC-MS
Mepanipyrim	anilinopyrimidine	FU	HPLC-MS
Metalaxyl	phenylamide	FU	HPLC-MS
Metalaxyl-M	phenylamide	FU	HPLC-MS
Metamitron	triazinone	HB	HPLC-MS
Metazachlor	chloroacetamide	HB	HPLC-MS
Methiocarb	carbamate	IN-MO-RE	HPLC-MS
Methiocarb sulfoxide (Methiocarb metabolite)	carbamate	IN-MO-RE	HPLC-MS
Methomyl	carbamate	IN-AC	HPLC-MS
Metolachor	chloroacetanilide	HB	HPLC-MS
Metribuzin	triazinone	HB	HPLC-MS
Myclobutanil	triazole	FU	GC-ECD
Nytempiram °	neonicotinoid	IN	HPLC-MS
Omethoate (Dimethoate metabolite) *	organophosphate	IN-AC	HPLC-MS
Oxamyl	carbamate	IN-AC-NE	HPLC-MS
Penconazole	triazole	FU	GC-ECD
Pencycuron	phenylurea	FU	HPLC-MS
Pendomethalin	dinitroaniline	HB	GC-ECD
Permethrin *	pyretroid	IN	GC-ECD
Phosmet	organophosphate	IN-AC	GC-ECD
Piperonyl butoxide °	safrole	SY	HPLC-MS
Pirimicarb	carbamate	IN	HPLC-MS
Pirimicarb-desmethyl (Pirimicarb metabolite)	carbamate	IN	HPLC-MS
Pirimiphos-methyl	organophosphate	IN-AC	HPLC-MS
Prochloraz	imidazole	FU	HPLC-MS
Procymidone *	dicarboximide	FU	GC-ECD
Propamocarb	carbamate	FU	HPLC-MS
Propiconazole *	triazole	FU	HPLC-MS
Propyzamide	benzamide	HB	HPLC-MS
Pyraclostrobin	strobilurin	FU	HPLC-MS
Pyridaben	pyridazinone	IN-AC	HPLC-MS
Pyrimethanil	anilinopyrimidine	FU	HPLC-MS
Pyriproxyfen	unclassified	IN	HPLC-MS
Quinoxyfen *	quinoline	FU	GC-ECD

Table A1. *Cont.*

Active Substances	Substance Group	Pesticide Type [a]	Detection Method
Rimsulfuron	sulfonylurea	HB	HPLC-MS
Rotenone *	isoflavones	IN	HPLC-MS
Spirodiclofen	tetronic acid	IN-AC	HPLC-MS
Spirotetramat	tetramic acid	IN	HPLC-MS
Tau-Fluvalinate	pyretroid	IN-AC	GC-ECD
Tebuconazole	triazole	FU	HPLC-MS
Tebufenozide	diacylhydrazine	IN	HPLC-MS
Tebufenpyrad	pyrazole	AC	HPLC-MS
Teflubenzuron	benzoylurea	IN	HPLC-MS
Tefluthrin	pyrethroid	IN	HPLC-MS
Tepraloxydim	cyclohexanedione	HB	HPLC-MS
Terbuthylazine	triazine	HB	HPLC-MS
Tetraconazole	triazole	FU	HPLC-MS
Tetramethrin *	pyrethroid	IN	GC-ECD
Thiabendazole	benzimidazole	FU	HPLC-MS
Thiacloprid	neonicotinoid	IN	HPLC-MS
Thiamethoxam	neonicotinoid	IN	HPLC-MS
Thifensulfuron-methyl	solfonylurea	HB	HPLC-MS
Thiobencarb *	thiocarbamate	HB	HPLC-MS
Thiodicarb *	carbamate	IN	HPLC-MS
Thiram	dithiocarbamate	FU	HPLC-MS
Thiophanate-methyl	benzimidazole	FU	HPLC-MS
Tolclofos-methyl	chlorophenyl	FU	GC-ECD
Tribenuron-methyl	solfonylurea	HB	HPLC-MS
Trifloxystrobin	strobulirin	FU	GC-ECD
Triflumuron	benzoylurea	IN	HPLC-MS
Triticonazole	triazole	FU	HPLC-MS

[a] AC, acaricide; BA, bactericide; FU, fungicide; HB, herbicide; IN, insecticide; MO, molluscicide; NE, nematicide; PG, Plant Growth regulator; RE, repellent; SY, synergist; * substance not approved in Italy; ° substance not classified in Italy.

References

1. Goulson, D.; Nicholls, E.; Botías, C.; Rotheray, E.L. Bee declines driven by combined stress from parasites, pesticides, and lack of flowers. *Science* **2015**, *347*, 1255957. [CrossRef] [PubMed]
2. Porrini, C.; Mutinelli, F.; Bortolotti, L.; Granato, A.; Laurenson, L.; Roberts, K.; Gallina, A.; Silvester, N.; Medrzycki, P.; Renzi, T.; et al. The status of honey bee health in Italy: Results from the nationwide bee monitoring network. *PLoS ONE* **2016**, *11*, e0155411. [CrossRef] [PubMed]
3. Gallai, N.; Salles, J.; Settele, J.; Vaissière, B.E. Economic valuation of the vulnerability of world agriculture confronted with pollinator decline. *Ecol. Econ.* **2009**, *68*, 810–821. [CrossRef]
4. Garibaldi, L.A.; Aizen, M.A.; Klein, A.M.; Cunningham, S.A.; Harder, L.D. Global growth and stability of agricultural yield decrease with pollinator dependence. *Proc. Natl. Acad. Sci. USA* **2011**, *108*, 5909–5914. [CrossRef]
5. Jacques, A.; Laurent, M.; EPILOBEE Consortium; Ribière-Chabert, M.; Saussac, M.; Bougeard, S.; Budge, G.E.; Hendrikx, P.; Chauzat, M.P. A pan-European epidemiological study reveals honey bee colony survival depends on beekeeper education and disease control. *PLoS ONE* **2017**, *12*, e0172591. [CrossRef]
6. Pham-Delègue, M.H. *Abeilles*; Editions de La Martinière: Paris, France, 1998; p. 47.
7. Lambert, O.; Piroux, M.; Puyo, S.; Thorin, C.; Larhantec, M.; Delbac, F.; Poliquen, H. Bees, honey and pollen as sentinels for lead environmental contamination. *Environ. Pollut.* **2012**, *170*, 33–43. [CrossRef]
8. Bonmatin, J.M.; Giorio, C.; Girolami, V.; Goulson, D.; Kreutzweiser, D.P.; Krupke, C.; Liess, M.; Long, E.; Marzaro, M.; Mitchell, E.A.; et al. Environmental fate and exposure; neonicotinoids and fipronil. *Environ. Sci. Poll. Res. Int.* **2015**, *22*, 35–67. [CrossRef]
9. Greatti, M.; Barbattini, R.; Stravisi, A.; Sabatini, A.G.; Rossi, S. Presence of the a.i. imidacloprid on vegetation near corn fields sown with Gaucho® dressed seeds. *Bull. Insectol.* **2006**, *59*, 99–103.

10. Rete Rurale. Available online: https://www.reterurale.it/flex/cm/pages/ServeBLOB.php/L/IT/IDPagina/1369 (accessed on 31 October 2019).

11. Porrini, C.; Sgolastra, F.; Renzi, T. Bee Emergency Service Team (BEST): Bee losses and mortality reports in Italy (2012–2014). *Bull. Insectol.* **2014**, *67*, 294.

12. Mutinelli, F.; Sgolastra, F.; Porrini, C.; Medrzycki, P.; Bortolotti, L.; Granato, A.; Gallina, A.; Lodesani, M. The status of honey bee health in Italy: Results from the nationwide monitoring network. In Proceedings of the XI European Congress of Entomology, Naples, Italy, 2–6 July 2018; p. 318.

13. EC. *Regulation (EC) No 1107/2009 of the European Parliament and of the Council of 21 October 2009 Concerning the Placing of Plant Protection Products on the Market and Repealing Council Directives 79/117/EEC and 91/414/EEC. OJ EU 2009, L 309*; EC: Brussels, Belgium, 2009; pp. 1–50. Available online: https://eur-lex.europa.eu/legal-content/EN/TXT/?uri=celex%3A32009R1107 (accessed on 31 October 2019).

14. EC. *Directive 2009/128/EC of the European Parliament and of the Council of 21 October 2009 Establishing a Framework for Community Action to Achieve the Sustainable Use of Pesticides. OJ EU 2009, L 309*; EC: Brussels, Belgium, 2009; pp. 71–86. Available online: https://eur-lex.europa.eu/legal-content/EN/ALL/?uri=celex%3A32009L0128 (accessed on 31 October 2019).

15. EC. *Commission Directive 2010/21/EU of 12 March 2010 Amending Annex I to Council Directive 91/414/EEC as Regards the Specific Provisions Relating to Clothianidin, Thiamethoxam, Fipronil and Imidacloprid. OJ EU 2010, L 65*; EC: Brussels, Belgium, 2010; pp. 27–30. Available online: https://eur-lex.europa.eu/eli/dir/2010/21/oj (accessed on 31 October 2019).

16. Italian ministry of Health. *Linee Guida Per la Gestione Delle Segnalazioni di Moria o Spopolamento Degli Alveari Connesse All'utilizzo di fitoFarmaci [Guideline for the Management of Reporting of Mortality or Dwindling of Beehives Linked to the Use of Pesticides] 0016168-31/07/2014-DGSAF-COD_UO-P*; Italian ministry of Health: Rome, Italy, 2014. Available online: http://www.izsvenezie.it/linee-guida-per-la-gestione-delle-segnalazioni-di-moriao-spopolamento-degli-alveari-connesse-allutilizzo-di-fitofarmaci (accessed on 31 October 2019).

17. Anastassiades, M.; Lehotay, S.J.; Stajnbaher, D.; Schenck, F.J. Fast and easy multiresidue method employing acetonitrile extraction/partitioning and 'dispersive SPE' for the determination of pesticide residues in produce. *JAOAC Int.* **2003**, *86*, 412–431. Available online: http://www.ingentaconnect.com/contentone/aoac/jaoac/2003/00000086/00000002/art00023 (accessed on 31 October 2019).

18. Lambert, O.; Piroux, M.; Puyo, S.; Thorin, C.; L'Hostis, M.; Wiest, L.; Bulete, A.; Delbac, F.; Pouliquen, H. Widespread occurrence of chemical residues in beehive matrices from apiaries located in different landscapes of western France. *PLoS ONE* **2013**, *8*, e67007. [CrossRef] [PubMed]

19. Calatayud-Vernich, P.; Calatayud, F.; Simò, E.; Picò, Y. Pesticide residues in honey bees, pollen and beeswax: Assessing beehive exposure. *Environ. Pollut.* **2018**, *241*, 106–114. [CrossRef] [PubMed]

20. Kiljanek, T.; Niewiadowska, A.; Gaweł, M.; Semeniuk, S.; Borzecka, M.; Posyniak, A.; Pohorecka, K. Multiple pesticide residues in live and poisoned honeybees—Preliminary exposure assessment. *Chemosphere* **2017**, *175*, 36–44. [CrossRef] [PubMed]

21. Wiest, L.; Bulete, A.; Giroud, B.; Fratta, C.; Amic, S.; Lambert, O.; Pouliquen, H.; Arnaudguilhem, C. Multi-residue analysis of 80 environmental contaminants in honeys, honeybees and pollens by one extraction procedure followed by liquid and gas chromatography coupled with mass spectrometric detection. *J. Chromatogr. A* **2011**, *1218*, 5743–5756. [CrossRef] [PubMed]

22. Mullin, C.A.; Frazier, M.; Frazier, J.L.; Ashcraft, S.; Simonds, R.; van Engelsdorp, D.; Pettis, J.S. High levels of miticides and agrochemicals in North American apiaries: Implications for honey bee health. *PLoS ONE* **2010**, *5*, e9754. [CrossRef] [PubMed]

23. EU. *Commission Implementing Regulation (EU) No. 485/2013 of 24 May 2013 Amending Implementing Regulation (EU) No 540/2011, as Regards the Conditions of Approval of the Active Substances Clothianidin, Thiamethoxam and Imidacloprid, and Prohibiting the Use and Sale of Seeds Treated with Plant Protection Products Containing Those Active Substances. OJ EU, 2013, L 139*; EU: Brussels, Belgium, 2013; pp. 12–26. Available online: http://eur-lex.europa.eu/LexUriServ/LexUriServ.do?uri=OJ:L:2013:139:0012:0026:EN:PDF (accessed on 31 October 2019).

24. EU. *Commission Implementing Regulation (EU) No. 781/2013 of 14 August 2013 Amending Implementing Regulation (EU) No 540/2011, as Regards the Conditions of Approval of the Active Substance Fipronil, and Prohibiting the Use and Sale of Seeds Treated with Plant Protection Products Containing this Active Substance. OJ EU 2013, L 219*; EU: Brussels, Belgium, 2013; pp. 22–25. Available online: http://eur-lex.europa.eu/LexUriServ/LexUriServ.do?uri=OJ:L:2013:219:0022:0025:EN:PDF (accessed on 31 October 2019).

25. Mitchell, E.A.D.; Mulhauser, B.; Mulot, M.; Mutabazi, A.; Glauser, G.; Aebi, A. A worldwide survey of neonicotinoids in honey. *Science* **2017**, *358*, 109–111. [CrossRef]
26. Calatayud-Vernich, P.; Calatayud, F.; Simó, E.; Pascual Aguilar, J.A.; Picó, Y. A two-year monitoring of pesticide hazard in-hive: High honey bee mortality rates during insecticide poisoning episodes in apiaries located near agricultural settings. *Chemosphere* **2019**, *232*, 471–480. [CrossRef]
27. Gaweł, M.; Kiljanek, T.; Niewiadowska, A.; Semeniuk, S.; Goliszek, M.; Burek, O.; Posyniak, A. Determination of neonicotinoids and 199 other pesticide residues in honey by liquid and gas chromatography coupled with tandem mass spectrometry. *Food Chem.* **2019**, *282*, 36–47. [CrossRef]
28. Simon-Delso, N.; Martin, G.S.; Bruneau, E.; minsart, L.-A.; Mouret, C.; Hautier, L. Honeybee colony disorder in crop areas: The role of pesticides and viruses. *PLoS ONE* **2014**, *9*, e103073. [CrossRef]
29. Christen, V.; Krebs, J.; Fent, K. Fungicides chlorothanolin, azoxystrobin and folpet induce transcriptional alterations in genes encoding enzymes involved in oxidative phosphorylation and metabolism in honey bees (*Apis mellifera*) at sublethal concentrations. *J. Hazard. Mater.* **2019**, *377*, 215–226. [CrossRef]
30. Johnson, R.M.; Dahlgren, L.; Siegfried, B.D.; Ellis, M.D. Acaricide, fungicide and drug interactions in honey bees (*Apis mellifera*). *PLoS ONE* **2013**, *8*, e54092. [CrossRef] [PubMed]
31. Sánchez-Bayo, F.; Goulson, D.; Pennacchio, F.; Nazzi, F.; Goka, K.; Desneux, N. Are bee diseases linked to pesticides?—A brief review. *Environ. Int.* **2016**, *89*, 7–11.
32. Sgolastra, F.; Medrzycki, P.; Bortolotti, L.; Renzi, M.T.; Tosi, S.; Bogo, G.; Teper, D.; Porrini, C.; Molowny-Horas, R.; Bosch, J. Synergistic mortality between a neonicotinoid insecticide and an ergosterol-biosynthesis-inhibiting fungicide in three bee species. *Pest Manag. Sci.* **2017**, *73*, 1236–1243. [CrossRef] [PubMed]
33. Martinello, M.; Baratto, C.; Manzinello, C.; Piva, E.; Borin, A.; Toson, M.; Granato, A.; Boniotti, M.B.; Gallina, A.; Mutinelli, F. Spring mortality in honey bees in northeastern Italy: Detection of pesticides and viruses in dead honey bees and other matrices. *J. Apic. Res.* **2017**, *56*, 239–254. [CrossRef]
34. Di Prisco, G.; Cavaliere, V.; Annoscia, D.; Varricchio, P.; Caprio, E.; Nazzi, F.; Gargiulo, G.; Pennacchio, F. Neonicotinoid clothianidin adversely affects insect immunity and promotes replication of a viral pathogen in honey bees. *Proc. Nat. Acad. Sci. USA* **2013**, *110*, 18466–18471. [CrossRef]
35. Chauzat, M.; Faucon, J. Pesticide residues in beeswax samples collected from honey bee colonies (*Apis mellifera* L.) in France. *Pest Manag. Sci.* **2007**, *63*, 1100–1106. [CrossRef]
36. EU Pesticides Database. Available online: https://ec.europa.eu/food/plant/pesticides/eu-pesticides-database/public/?event=homepage&language=EN (accessed on 19 December 2019).
37. Kochansky, J.; Wilzer, K.; Feldlaufer, M. Comparison of the transfer of coumaphos from beeswax into syrup and honey. *Apidologie* **2001**, *32*, 119–125. [CrossRef]
38. Mutinelli, F. Veterinary medicinal products to control Varroa destructor in honey bee colonies (*Apis mellifera*) and related EU legislation—An update. *J. Apic. Res.* **2016**, *55*, 78–88. [CrossRef]
39. EU. *Regulation (EU) No 528/2012 of the European Parliament and of the Council of 22 May 2012 Concerning the Making Available on the Market and Use of Biocidal Products*. OJ EU, 2012, L167; EU: Brussels, Belgium, 2012; pp. 1–123. Available online: https://eur-lex.europa.eu/legal-content/EN/TXT/?uri=CELEX%3A32012R0528 (accessed on 31 October 2019).
40. PPDB. Pesticide Properties Database. University of Hertfordshire. 2016. Available online: http://sitem.herts.ac.uk/aeru/ppdb/en/index.htm (accessed on 31 October 2019).
41. Johnson, R.M.; Pollock, H.S.; Berenbaum, M.R. Synergistic interactions between in-hive miticides in *Apis mellifera*. *J. Econ. Entomol.* **2009**, *102*, 474–479. [CrossRef]
42. Pilling, E.D.; Jepson, P.C. Synergism between EBI fungicides and a pyrethroid insecticide in the honeybee (*Apis mellifera*). *Pestic. Sci.* **1993**, *39*, 293–297. [CrossRef]

 *diversity*

Article

Control of *Varroa destructor* Mite Infestations at Experimental Apiaries Situated in Croatia

Ivana Tlak Gajger [1,*]**, Lidija Svečnjak** [2]**, Dragan Bubalo** [2] **and Tomislav Žorat** [3]

[1] Department for Biology and Pathology of Fish and Bees, Faculty of Veterinary Medicine, University of Zagreb, 10000 Zagreb, Croatia

[2] Department of Fisheries, Apiculture, Wildlife Management and Special Zoology, Faculty of Agriculture, University of Zagreb, 10000 Zagreb, Croatia; lsvecnjak@agr.hr (L.S.); dbubalo@agr.hr (D.B.)

[3] Veterinary ambulance Obrovac d.d., 23450 Obrovac, Croatia; tomislav.zorat@gmail.com

* Correspondence: ivana.tlak@vef.hr; Tel.: +385-91-2390-041

Received: 5 December 2019; Accepted: 22 December 2019; Published: 25 December 2019

Abstract: Experimental varroacidal treatments of honey bee colonies were conducted on five apiaries (EA1–EA5) situated at five different geographical and climatic locations across Croatia. The aim of this study was to assess the comparative efficacy of CheckMite+ (Bayer, Germany), Apiguard (Vita Europe Ltd.; England), Bayvarol, C, (Bayer, Germany), Thymovar, (Andrma BioVet GmbH, Germany), and ApiLife Var, (Chemicals Laif SPA; Vigonza, Italy) for controlling the honey bee obligatory parasitic mite *Varroa destructor* in different conditions in the field during summer treatment. The relative varroa mite mortality after treatments with applied veterinary medicinal products were EA1 (59.24%), EA2 (47.31%), EA3 (36.75%), EA4 (48.33%), and EA5 (16.78%). Comparing the relative efficacy of applied varroacides, the best effect was achieved with CheckMite+, and the lowest for honey bee colonies treated with Apiguard (statistically significant difference was confirmed; $p < 0.05$). Considering the lower efficacy of thymol-based veterinary medicinal products observed on all EA in these study conditions, it may be concluded that their use is limited under different treatment regimes. Despite unfavourable weather and environmental conditions, with exceptions of EA5/EA5′ and EA1, the relative varroacidal efficacy of authorized veterinary medicinal product treatments in moderately infested honey bee colonies ensured normal overwintering and colony development during next spring.

Keywords: *Apis mellifera*; *Varroa destructor*; experimental apiaries; varroacidal efficacy; VMP

1. Introduction

Honey bee colonies are valuable ecological and economical insects. They are fundamental for the food production and ecosystem biodiversity through their important role in pollination of cultivated crops and wild plants [1,2]. Different adverse environmental conditions, including pathogens and pests, have been implicated in the dissipated health of honey bee colonies or even their losses [3]. Although causes of extensive honey bee colony losses appear to be multifactorial [4], the obligate ectoparasitic mite *Varroa destructor* of honey bees is a major contributing factor [5]. As the *V. destructor* mite feeds on the haemolymph and fat body of adults [6] and developing stages of honey bees, and additionally facilitates the transmission of certain viruses [7–9], this seems to have a significant negative effect on the host immune response [10]. Because the aforementioned possible consequences of mite infestation consist in damages on the individual and colony level, the *V. destructor* mite population in honey bee colonies requires regular control during the whole year.

Moreover, honey bee colonies are highly influenced by beekeepers' management practices, socioeconomic conditions, and the level of implementation of policies supporting beekeeping activities [11] in the particular country. In view of the spread of varroosis across Europe and the problems which this disease has brought in the beekeeping sector, the European Union (EU) has been encouraged to set up national programs aimed at improving the general conditions for the sustainable beekeeping production and management in its ecological, economic, and social dimensions [12].

In Croatia, as a new member of the EU, such apicultural programs have existed for several years. These national beekeeping programs (2011–2013, before accessing EU; 2014–2016; 2017–2019) employed a high participation rate of beekeepers [13] but failed to a certain extent in terms of the efficacy of the varroosis control by using authorized veterinary medicinal products (VMP). This is followed by the beekeepers' explanation that there are very few acaricide active substances on the market [14]. This is evident from the fact that in the past, beekeepers have not made any kind of notes or records of evidence regarding the use of varroacides as active ingredients and its efficacy on *V. destructor* mites. Over a few consecutive years, there has been a visible decreasing trend in procuration of authorized VMPs for usage in beekeeping [14], despite their total number, as well as the number of active substances, increasing by more than double [15].

According to the Varroa Control Program, which is a part of the Croatian National Regulation on animal protection measures against infectious and parasitic diseases and related financing from 2011 till today (with annual minor modifications) [16], beekeepers nationwide are required to implement one obligatory treatment of honey bee colonies against *V. destructor* mites using an authorized VMP in combination with other measures of integrated pest management implemented in an appropriate time of year [17]. Application of VMPs is recommended in the period from 1 July to 31 August, after main honey harvesting, and with the schedule dependent on the geographical, climatic and honey bee forage conditions in different regions. The main model is also to conduct treatments on all colonies and apiaries in the same area and during the same period. In this way, reinfestation and horizontal mite transfer between apiaries during robbing behavior [18,19] can be avoided.

Experimental apiaries (EA) in Croatia were established in 2013 with the purpose of conducting different research studies with a possible comparison of results linked to different environmental impacts.

There are numerous active acaricidal substances available incorporated into different formulations of medicines, different methods of application, and techniques for *V. destructor* mite population control. A variety of synthetic varroacides have been widely used in the last few decades with variable effects, due to geographical regions or often in response to bad beekeeping practices, such as multiple consecutive and repeated use of the same acaricide, sub- or overdosage, too short or prolonged treatment duration, improper time and way of acaricidal product application, too few active substances in the same time treatment, etc., which has led to increased tolerance to most of them. *V. destructor* mites have developed resistance to the most widely used synthetic varroacides [20–22]. In order to avoid the accumulation of chemical residues in honey and beeswax, beekeepers are very interested in treatments of natural, ecofriendly, so-called *soft* acaricides, which are often inconsistent, more variable, or effectiveness or therapeutically limited under different climatic conditions [23–26].

The aim of this study was to evaluate and compare the varroacidal efficacy and mite mortality dynamic during summer treatment of honey bee colonies situated at EA influenced by different environmental conditions, treated with five different available authorized VMPs (CheckMite+, Apiquard, Bayvarol, Thymovar, and ApiLife Var) used simultaneously and compared with negative control (untreated group of honey bee colonies). The percentages of *V. destructor* mite mortality by the experimental treatments were estimated according to recommendations of the European Medicines Agency (EMA) [27]. Efficacy of treatments was also compared between EA. Additionally, the commonly used oxalic acid was used for follow-up winter treatment in broodless honey bee colonies to establish the final parasitic mite drop.

2. Materials and Methods

2.1. Locations of Experimental Apiaries and Field Trail Design

The study was conducted during active beekeeping season of 2014 at five different apiaries: EA1–EA5 (Figure 1; http://geoportal.dgu.hr/) in Croatia. EAs were located in different geographical regions: EA1 (46° 5′ 37″ N, 15° 53′ 34″ E); EA2 (45° 48′ 16″ N, 18° 39′ 54″ E); EA3 (45° 13′ 45″ N, 13° 56′ 29″ E); EA4 (44° 36′ 54″ N, 43° 55′ 46″ E), and EA5 (43° 55′ 47″ N, 16° 26′ 18″ E). EA1 was situated in grassland surrounded by fields where intensive agriculture practice is in use; EA2 was on the lea surrounded with vegatable gardens; EA3 was located in a rural area surrounded with orchards and vineyards; EA4 was in the deciduous wood, and EA5 was situated on the grassland prairie. Due to the lack of natural food during July and August on the EA5 location, the honeybee colonies were moved after summer treatment against *V. destructor* mites to a new, more favourable location on the Adriatic sea island Vis. The new position of this apiary was annotated as EA5′ (43° 2′ 45″ N, 16° 9′ 14″ E).

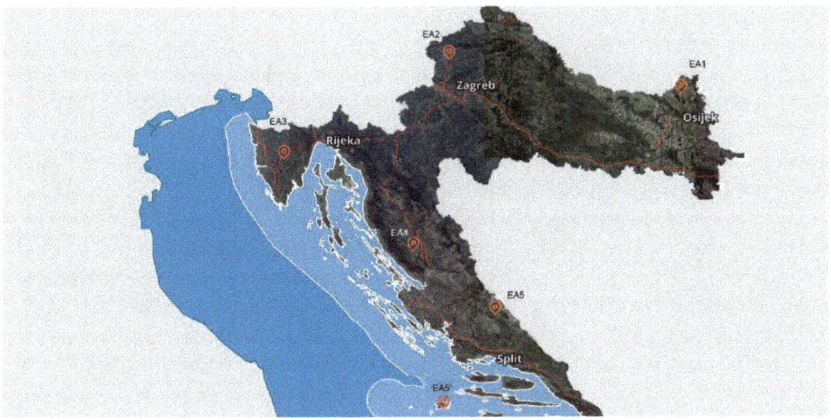

Figure 1. Locations of experimental apiaries in Croatia.

Each EA consisted of 30 honey bee colonies (*Apis mellifera carnica* Pollmann, 1879) placed in standard Langsthrot Root (LR) hives. Experimental colonies were queen right, had combs occupied with adult honey bees, and were fully developed and productive. Prior to the experimental period, all experimental honey bee colonies were uniformed in respect to brood size, the comb area covered with adult bees, and amount of stored food [28]. Honey bee queens were one year old. Experimental colonies were also visually inspected for the presence of pathology signs on adult bees and brood. Adult bees showed normal behavior, and there were no visible signs of infectious diseases on brood. No acaricidial treatment of the honey bee colonies was done prior to the start of experimental treatments. After inspection, the honeybee colonies situated on EA1–EA5 were divided into six experimental groups (A, B, C, D, E, and O), and each group consisted of five beehives (Table 1).

All beehives were equipped with varroa mite screen boards for monitoring mite fall counts. In early spring, metal sheets were placed on the bottom board of each bee hive in order to record the natural mite mortality prior to treatments, and later in the season, the mite drop after the experimental treatments. Above the sheets, wire screens were installed to prevent contact of the adult bees with debris and to prevent ants from removing dropped *V. destructor* mites. Experimental and control colonies were monitored for mites mortality in prior, during, and after treatments performed in brood and broodless (winter) periods. Mite counts were carried out every day during summer treatment (A, C, D, and O groups—42 reads; B and E groups—28 reads were recorded for each colony), and seven days after winter treatment. At EA5′, the emergency autumn treatment was performed on 16 September by the commonly used amitraz on a one-time basis.

Table 1. Field trail design.

Apiary	Pretreatment Mite Fall $n = 150$	CheckMite+ A $n =$ 5/apiary	Apiguard B $n =$ 5/apiary	Bayvarol C $n =$ 5/apiary	Thymovar D $n =$ 5/apiary	ApiLife Var E $n =$ 5/apiary	Oxalic Acid A,B,C,D,E,O
EA1	1.6–24.7	25.7–5.9	25.7–22.8	25.7–5.9	25.7–5.9	25.7–22.8	28.11–6.12
EA2	1.6–17.7	17.7–28.8	17.7–14.8	17.7–28.8	17.7–28.8	17.7–14.8	29.11–7.12
EA3	1.6–15.7	16.7–27.8	16.7–13.8	16.7–27.8	16.7–27.8	16.7–13.8	11.12–18.12
EA4	1.6.–16.7	17.7–28.8	17.7–14.8	17.7–28.8	17.7–28.8	17.7–14.8	1.12–8.12
EA5	1.6–14.7	15.7–26.8	15.7–12.8	15.7–26.8	15.7–26.8	15.7–12.8	–

Note: On every experimental apiary (EA), the negative control was included (untreated group of honeybee colonies, O; n = 5/apiary); treatment of broodless honeybee colonies with oxalic acid was carried out on all survivals during the winter period, except in EA5.

2.2. Drugs and Treatments

Treatments were conducted during the summer season immediately after the main honey harvesting. The authorized VMPs were used in the recommended doses. CheckMite+ (Bayer, Leverkusen, Germany), based on coumaphos as the active ingredient, was applied in the form of two beehive pest control strips inserted between frames, with waxcombs sealed with honeybee brood in the brood chamber for a 42 day period. Apiguard (Vita Europe Ltd.; Basingstoke, England) is an authorized thymol-based acaricide packed in an aluminum tray, so its coated sheets were placed on the top bars of the bee hive frames of each brood chamber, one tray during two weeks and the second one during the consecutive two weeks. The treatment with flumethrin in an authorized VMP—Bayvarol (Bayer)—was applied as four pest control strips were inserted between frames with sealed honey bee brood in the brood hive chamber for a 42 day period. Thymovar (Andermatt BioVet GmbH, Lörrach, Germany), formulated on cellulose wafer, contains 15 g of thymol and was used as one piece cut in two parts, which were separately placed on the top bars of frames on a two-time-basis for 42 days in total. ApiLife Var (Chemicals Laif SPA; Vigonza, Italy) is based on a few active ingredients (thymol, eucalyptus oil, levomenthol, and camphor) imbibed in vermiculite tablets. One tablet per bee hive was applied every eight days with four applications in total. Every portion of the tablet was placed in a corner of the brood hive chamber and remained in the honey bee colonies to be chewed and removed by the adult bees.

The oxalic acid solution was prepared using 1 L of sugar syrup to dissolve 35 g of oxalic acid dehydrate (Kemika, Zagreb, Croatia). The sugar syrup was prepared by mixing hot freshwater (70–80 °C) with commercial sugar (Viro, Virovitica, Croatia) (1:1). Prepared oxalic acid solution was administered to the honey bee colonies cold with a syringe trickling 5 mL for each intercomb space occupied by adult bees, from the top. The number of phoretic mites fallen after the winter treatment was counted by using metal label sheets on the bottom boards, checked every day for one week.

For autumn treatment, Varidol (TolnAgro Kft., Szekszárd, Hungary) was used once, in honeybee colonies at location EA5′, by fumigation according to instructions for use.

2.3. Meteorological Conditions

Data on weather conditions (air temperature (°C), relative air moisture (%), number of days with rain; average values per month) during monitored beekeeping season were obtained from Croatian Hydrometeorological Department, from local climatic-meteorological stations (Osijek, Krapina, Pazin, Gospić, Sinj). All parameters were measured three times per day at 7:00 a.m., 2:00 p.m., and 9:00 p.m.

2.4. Estimating the Strength of Honeybee Colonies

To estimate the strength of honey bee colonies, a Liebefeld method was performed, with visual determination of number of adult honey bees and brood sealed in beeswax combs [28]. The estimation of honey bee colonies was conducted three times (prior—I, during—II, and after experimental summer treatments—III), during morning hours, between 9:00 and 10:00 a.m. before the first massive forage flights of honey bees. Strength of honey bee colonies was estimated as follows: EA1—19 May, 25 July, and 16 September; EA2—29 April, 17 July, and 8 September; EA3—14 May, 16 July, and 7 September; EA4—20 May, 10 July, and 2 September; and EA5—28 April, 15 July, and 10 September. Owing to easier assessment with adult honey bees or brood-covered comb areas, the frame for the LR hive was used and prior divided with a plastic grid into 1 dm^2 quadrants.

2.5. Colony Examinations, Mite Counts and Treatments Efficacy

Clinical examination of honeybee colonies included visual inspection of honeybee brood and adult bees, the behaviour of adult bees, as well as activity of bees at the entrances of hives. Fallen *V. destructor* mites were counted during the pretreatment, treatment, and a particular number of days after each treatment, and the sum of those results calculated after the final treatment was considered the total mite drop. The proportion of mites falling after each treatment to the total number of fallen mites was estimated in percentages (%). The efficiency of each experimental treatment (A, B, C, D, E) was estimated according to the recommendations of the EMA [27].

2.6. Data Analysis

The data analyses were performed by one-way analysis of variance (ANOVA) using statistical software package Statistica-StatSoft v.7. (StatSoft, Inc, Tulsa, OK, U.S.A.). The results were presented as the mean values, standard deviations, and standard errors. To assess the statistical differences in honeybee colonies' strength (estimated number of honeybees per colony), they were compared between groups and three estimation dates (April/May, July, and September), and the number of fallen *V. destructor* mites between experimental groups (A, B, C, D, E) and control group (O) at different locations and estimation dates, the Kruskal–Wallis test was performed (α = 0.05). Multisample comparison using Kruskal–Wallis H test was carried out to compare the mean values of the number of fallen *V. destructor* mites between experimental and control groups in the pretreatment, treatment, and after-treatment period.

3. Results

3.1. Meteorological Conditions

The average values of air temperatures, relative air moisture, and number of days with rain for each location of EA during the observed period of beekeeping season 2014 is shown in Table 2. Obviously, these harms encountered in beekeeping may indicate that this acaricidal therapy of honey bee colonies with authorised VMPs had the obtained effect instead of especially detrimental environmental circumstances.

Table 2. Meteorological conditions at different apiaries.

| | Meteorological Circumstances (Average Values Per Month) | | | | | | | | | | | | | |
| | Air Temperature (°C) | | | | | Relative Air Moisture (%) | | | | | Number of Days with Rain | | | | |
Month	EA1	EA2	EA3	EA4	EA5	EA1	EA2	EA3	EA4	EA5	EA1	EA2	EA3	EA4	EA5
Mar	9.5	9.9	8.9	7.0	9.5	74	65	67	69	68	9	8	5	10	9
Apr	13.2	12.9	11.9	10.5	12.6	74	73	74	72	71	11	12	8	10	15
May	16.1	14.9	14.4	12.9	15.1	73	69	72	66	69	11	18	3	7	14
Jun	20.5	19.2	19.6	17.7	20.1	67	74	66	67	66	3	9	2	4	10
Jul	21.8	20.7	20.0	18.6	21.1	74	76	78	73	73	3	10	6	10	12
Aug	20.8	19.1	19.4	18.3	20.9	76	79	79	72	73	6	11	8	2	7
Sep	17.0	15.6	15.7	13.7	16.7	82	86	81	84	79	10	17	11	13	14
Oct	13.3	12.8	13.0	11.2	13.7	82	85	83	79	76	9	9	8	7	10
Nov	8.3	8.6	11.0	8.5	10.9	87	89	87	86	84	3	17	16	13	14

3.2. *Estimating the Strength of Honeybee Colonies*

Despite the equalization of honeybee colonies before pretreatment period with respect to colony strength, some statistical differences were determined between experimantal and control groups of honey bee colonies at different estimation days (I, II, or III) at EA1, EA2, and EA5, as follows: EA1: I ($p < 0.001$; F = 9.87), II ($p < 0.05$; F = 3.32), III ($p < 0.01$; F = 3.94); E2: III ($p < 0.001$; F = 9.87); and E5: III ($p < 0.05$; F = 3.55). Variations in the average number of honey bees per group during three estimation terms are shown in Figure 2.

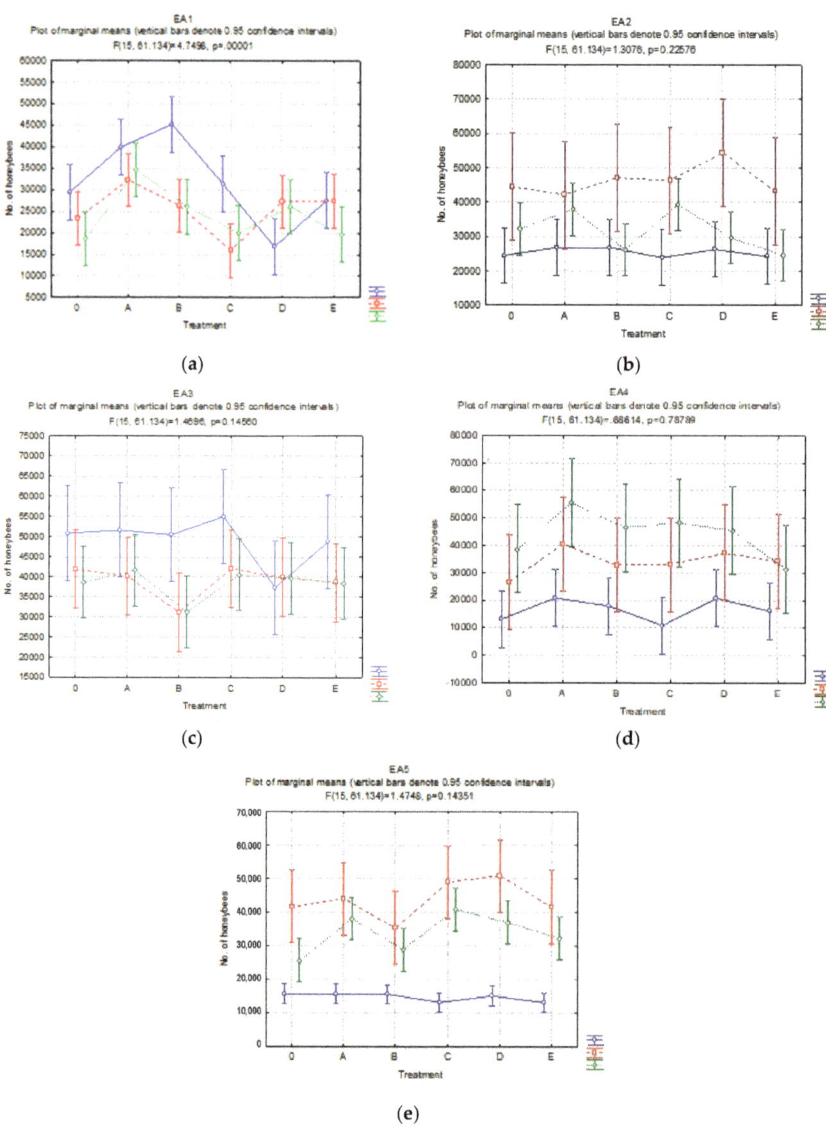

Figure 2. Honey bee colony strength differences between experimental and control groups by different estimation dates (I—blue lines, II—red lines, III—green lines) at five apiary locations: (**a**) EA1, (**b**) EA2, (**c**) EA3, (**d**) EA4, (**e**) EA5.

3.3. V. destructor Mite Fall Prior to, during, and after Varroacidal Treatments

3.3.1. Pretreatment Period

During the pretreatment periods the average daily mite drops in 30 honey bee colonies per apiary were: EA1, 2.76 (± 2.60); EA2, 3.04 (± 2.60); EA3, 0.80 (± 0.10); EA4, 4.32 (± 0.50); and EA5, 0.09 (± 0.10). These values did not differ significantly between the experimental groups on individual apiaries but were significantly different between EA locations (EA4, EA5; $p < 0.001$). Results are presented in Figure 3.

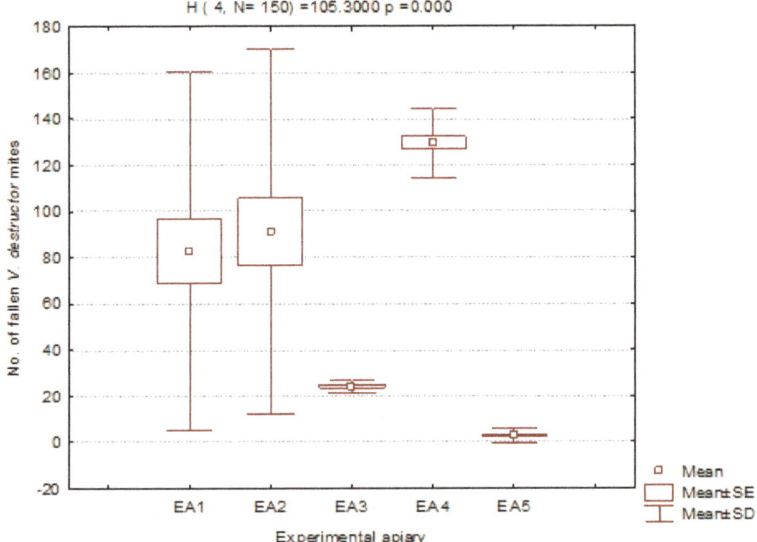

Figure 3. Natural *V. destructor* mite drop during the pretreatment period at five apiary locations (EA1–EA5).

3.3.2. Treatment Period

During the summer treatment period, the estimated total *V. destructor* mite mortality that resulted was significantly higher than natural mite drop. Results of summer and winter treatment (to determine the residual amount of mites) of honey bee colonies with VMPs are shown in Figure 4. Because of very low efficacy of summer treatments in all groups of experimental colonies situated on EA5 (Figure 5), the necessary follow-up autumn treatment was done at EA5' (Figure 5), and calculations of treatments efficacy was based on those mite drops (A—47.52%; B—3.11%; C—25%; D —5.18%; E—3.11%). Results of this EA were analyzed separately from those of other EAs.

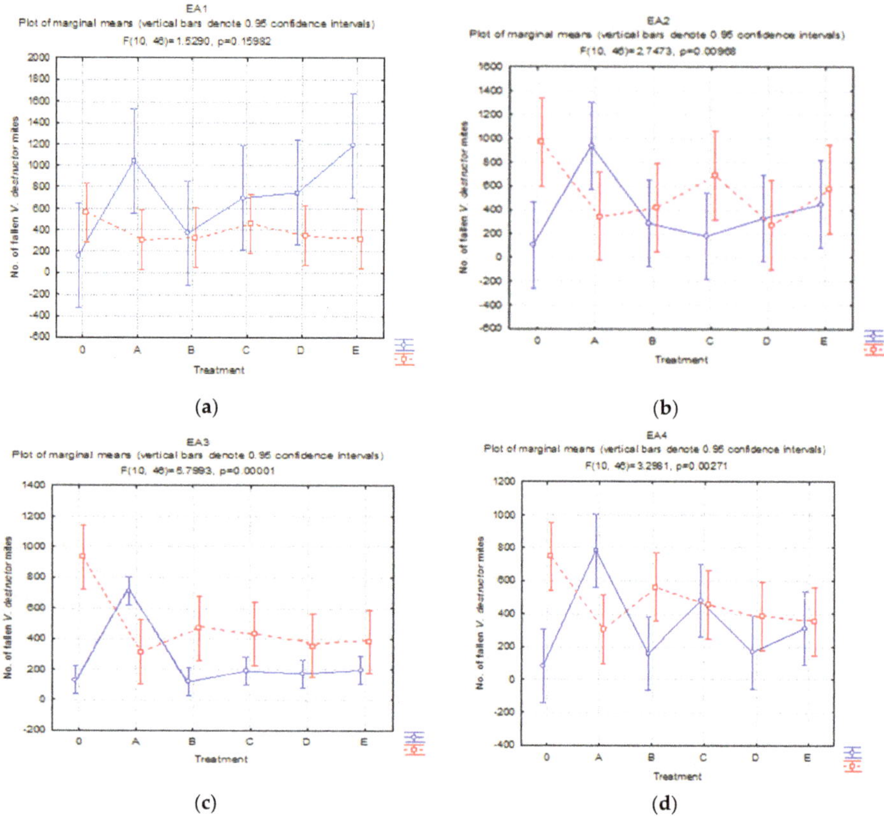

(a) (b)

(c) (d)

Figure 4. Differences in the number of fallen *Varroa destructor* mites between experimental groups (A, B, C, D, E) and control groups (O), during summer (blue lines) and winter treatment (red lines); mean ± SD. (**a**) EA1, (**b**) EA2, (**c**) EA3, (**d**) EA4.

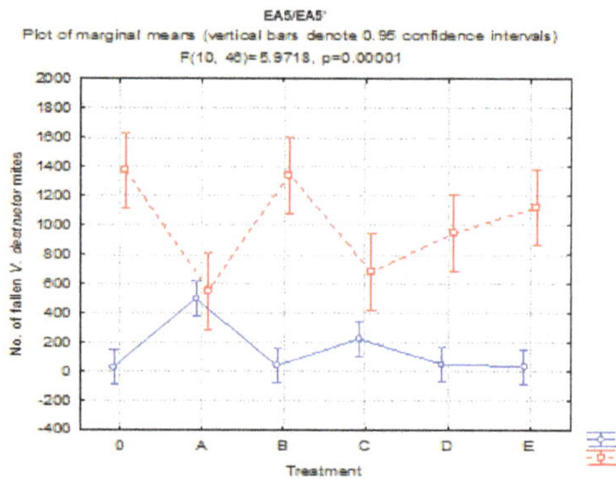

Figure 5. Differences in the number of fallen *V. destructor* mites between experimental groups (A, B, C, D, E) and control groups (O), during summer (blue lines) and follow-up autumn treatment (red lines), at EA5/EA5′; mean ± SD.

The mean values of *V. destructor* mortality on EA1 did not differ significantly between the experimental groups (A, B, C, D, E), but it was different in comparison with the control group ($p < 0.05$; h = 14.2). Summer treatment results were significantly different from those of winter treatment ($p < 0.05$; F = 2.74). Efficacy of varoacidal treatment is presented in Table 3, and it was decreased as follows: A > E > D > C > B. Prior to winter treatment, nine of the treated honey bee colonies were dead: 3 colonies from B group, 3 colonies from E group, 2 colonies from D group and 1 colony from C group.

Table 3. Treatment efficacy of varroacides on different experimental apiaries.

EA	Experimental Group	Number of Fallen *V. destructor* Mites (S/W)			% of Treatment Efficacy
-	-	Mean	Min.	Max.	
EA1	A	1044/286.4	430/237	1967/449	78.50
	B ‡‡‡	371.6/660.5	149/616	975/705	36.00
	C ‡	699.2/546	168/490	1112/609	56.20
	D‡‡	780/509.33	283/357	1605/774	60.50
	E ‡‡‡	1188.8/642	429/604	2046/680	65.00
EA2	A	950/326.6	381/135	2217/671	74.42
	B	287.8/421.2	65/177	744/658	40.60
	C	200.4/680.8	96/111	326/1038	22.75
	D	332.2/273.6	152/85	664/638	54.90
	E	448.4/572.8	48/100	1177/1195	43.90
EA3	A	724/314	603/138	802/435	69.75
	B	118.4/489.2	97/233	159/635	19.49
	C	190.4/433.4	115/231	346/625	30.52
	D	171.6/379.6	114/168	333/589	31.13
	E	194.6/397	102/168	442/602	32.90
EA4	A	783.2/264.4	125/278	1207/382	74.77
	B	195/542.4	43/384	381/789	26.45
	C	480.2/405.2	279/240	766/563	54.24
	D	167.2/406.6	48/180	373/594	41.13
	E ‡	312.6/380.7	30/126	1119/594	45.09

Note: S—summer treatment; W—winter treatment; ‡, ‡‡, ‡‡‡—number of dead honey bee colonies prior to winter treatment, in particular experimental groups.

At the location of EA2, each treatment (A, B, C, D, E) induced a significantly higher ($p < 0.05$; h = 12.9) *V. destructor* mite mortality in the parallel untreated control colonies (O). The summer treatments and winter treatments ($p < 0.005$; F = 2.83) were also statistically significantly different. Efficacy of different VMP treatments (Table 3) decreased as follows: A > D > E > B > C. All honey bee colonies survived.

The mortality rates in the treated colonies of EA3 apiary were significantly different from those in the control honey bee colonies ($p < 0.05$; h = 12.77), and between the summer and winter treatments ($p < 0.001$; F = 5.79). Efficacy of acaricidal treatments (Table 3) decreased in the same order as on the EA1.

Total V. destructor mite drop after the treatments were significantly higher ($p < 0.005$; h = 16.6) than number of dead mites counted on the bottom boards hive inserts of untreated control honey bee colonies situated at EA4. Also, the differences between experimental groups were determinated ($p < 0.01$; F = 5.95), during the summer treatments. Order of VMPs treatment efficacy decreased as follows: A, C, E, D, and B.

Varroacidal efficacy of applied VMPs was significantly different for treatments of experimental honey bee colonies from groups A and B (Figure 6a), at each EA ($p < 0.001$, F = 6.933). As a consequence of nontreated honey bee colonies during the summer period, all control groups were also significatly lower regarding percentages of varroacidal efficacy. Overall varroacidal efficacy at individual EA did not differ significantly between EA1, EA2, EA3, and EA4. These results are presented in Figure 6b.

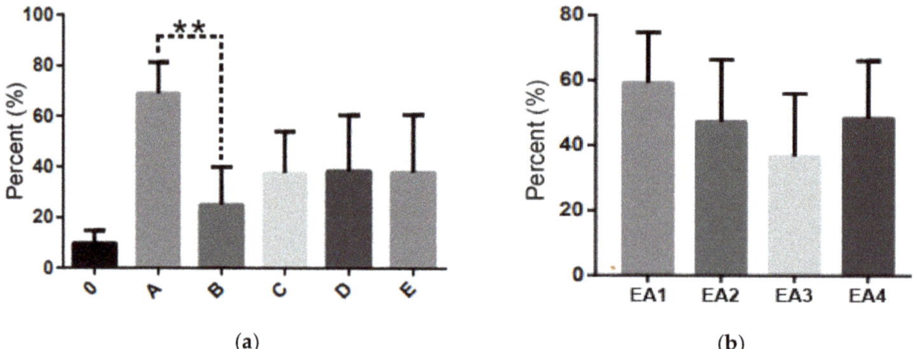

(a)   (b)

Figure 6. Treatment efficacy of varroacides with different active ingredients; (**a**) treatments; (**b**) locations; ** ($p < 0.05$); mean ± SD.

4. Discussion

Varroa mites, acting simultaneously with other stressors, are known to be a main cause of honey bee colony weakening and collapsing [5,29]. The *V. destructor* mite population requires regular control, correctly and in a timely manner, of honey bee colony management. Five different VMPs, only authorized in Croatia in 2014, were tested at five apiaries to test their performance in different field conditions because control of varroosis by using VMPs is widely acknowledged to be an essential part of beehive management [30].

The 2014 active beekeeping season was generally very weak in different aspects. There were more days with rain, very low average year air temperatures, along with a mild winter, and consequently, a very early honey bee brood development was noticed, even in January. As one replication cycle of *V. destructor* mites is about 13 days, multiple replications of mites can be expected to have occurred [17]. Weather circumstances were also very favorable for a fast-increasing growth trend of *V. destructor* mite populations in honey bee colonies with early reared brood [31]. Additionally, lack of natural food in the environment effected a decreasing of immunological status of the honey bee colonies and induced stress caused by hunger. All these detrimental impacts of environmental factors, high *V. destructor* mite infestations, and weak main pastures ended very early, so in most regions, the first summer varroacidal treatment was essential.

A different number of *V. destructor* mite drops was detected during the pretreatment period, which was probably a consequence of a different level of parasitic colony invasions at different EAs. Meeting the demand of no varroacidal treatments that year, before the start of the described study means that they were treated the last time in the winter preceding the experimental period. This may have implicated a situation where establishing new summer generations in all honey bee colonies coincided with a high level of *V. destructor* mite infestation, which may have damaged colonies in each experimental group. Similar findings were published before [32].

Because of extremely low amounts of stored honey in hives, at EA2 and EA5, the honey extraction was not done for the whole year.

Treatments against varroa mites started at different dates for each EA because the blooming of plants that make up the main bee pastures finished at different times. The best relative varroacidal efficacy was seemingly achieved at EA1 (59.24%), but if an analysis of the survival of treated colonies is included, then it is clear that 30% of them collapsed before the next active beekeeping season. Most of the lost honey bee colonies were treated with VMPs based on thymol as active ingredients. Here it must be stressed that first summer treatments on EA1 started only towards the end of July, which is relatively late but still in accordance with advised varroosis control schedules in Croatia.

Although relative average treatment efficacy at EA2 (47.31%), EA3 (36.75%), and EA4 (48.33%) was pretty low and below expectations, all honey bee colonies survived and successfully overwintered (except one honey bee colony at EA4, treated with a thymol-based VMP).

Comparing the relative efficacy of the used varroacides with different active ingredients (Figure 6a), the best effect was achieved for honey bee colonies from an A experimental group, and lowest for colonies from a B experimental group, where statistically significant difference was affirmed ($p < 0.05$; F = 6.93). Although the organic acaricides have certain advantages after repeated use, their efficacy may be inconsistent and more variable compared with synthetic acaricide formulations [24,33–36], which is also confirmed with our results.

Changes in honey bee colonies' strength trends were different between apiary locations, but within the expected ranges under the study and environmental conditions, as well as beekeeping practices (EA1, EA2, EA3, and EA4).

At location EA5 at the beginning of active beekeeping season, honey bee colonies were very weak with low numbers of adult bees (Figure 2). Then, in the middle of the active season, there was an opposite situation: a high number of adult bees without brood or very few comb cells were sealed with development stages of bees. At the same time there was non food in nature, and honey bees ate away almost all food storages in their hives. After emergency varroacidal treatment, which produced higher efficacy compared to earlier treatments because larger proportion of phoretic mites being on adult bees [37], and transport at island Vis (EA5′), they came in with much better environmental circumstances with plenty of natural food. In the next few weeks, the strength of colonies increased quickly, but because of serious damages, bees got high varroa infestation and insufficient efficacy of VMPs treatment during the summer treatment (except A group—47.52%), most of the colonies from other treated groups died before the next spring (63.4%). The efficacy of the VMP treatments was under expectations, probably due to fast reproduction of the surviving mites, but also because of possible reinfestations during experiments [35].

5. Conclusions

Varroa mite infestations control using CheckMite+ (Bayer, Germany), Apiguard (Vita Europe Ltd.; England), Bayvarol (Bayer, Germany), Thymovar (Andrma BioVet GmbH, Germany), and ApiLife Var (Chemicals Laif SPA; Vigonza, Italy) at five different EAs in Croatia during the beekeeping season of 2014 induced a higher mite mortality compared to control, as well as in comparison with mite drop in the pretreatment period. Despite unfavourable weather and environmental conditions, with the exceptions of EA5/EA5′ and EA1, the relative varroacidal efficacy of authorized VMP treatments in moderately infested colonies ensured normal overwintering and colony development during the next spring. Due to a lower efficacy of thymol-based VMPs observed at all EAs in this study conditions, it may be concluded that their use is limited under different treatment regimes. The results of this study imply that efficacy of used varroacidals strongly depends on geography, but also on timely manner, and can vary from season to season.

It can be concluded that an adequate *V. destructor* mite control must include a few measures, primarily good beekeeping maintenance techniques in combination with appropriate use of authorized VMPs. Different treatment regimes should also be applied with continuous parasitic mite mortality monitoring. Application of varroacides should be performed after the main honey flow, on all apiaries of the same epizootiology area, and in all honey bee colonies with mite infestations levels above the economic threshold. The same and appropriate treatment timing will ensure honey bee colonies surviving and prevent reinfestations. In specific situations, it is possible to use emergency treatments and alternate synthetic acaricides with food additives with acaricidal effect in rotation programs in order to decelerate the resistance of varroa mites to multiply used acaricides and to reduce the impact of increasing comb wax contamination. All varroacidal treatments must be performed in accordance and in combination with other specific regulations ordered by the national authorities.

Author Contributions: Conceptualization, I.T.G.; methodology, I.T.G.; formal analysis I.T.G. and L.S.; investigation, I.T.G., L.S., D.B., and T.Ž.; resources, I.T.G. and D.B.; data curation, L.S. and I.T.G.; writing—original draft preparation, I.T.G.; writing—review and editing, L.S., D.B., T.Ž., and I.T.G.; visualization, L.S. and I.T.G. All authors have read and agreed to the published version of the manuscript.

Funding: This research was funded by Project—Impact of diseases epidemiology and environmental factors to the biological-rearing status of honey bee colonies; Croatian National Beekeeping Program 2014.

Acknowledgments: The authors would like to thank colleagues Saša Prđun and Josipa Vlainić for technical assistance; and beekeepers Dario Dješka, Ranko Anđelini, Ivan Balenović, and Antionio Mravak for technical support and maintenance of the experimental apiaries.

Conflicts of Interest: The authors declare no conflict of interest. The funders had no role in the design of the study; in the collection, analyses, or interpretation of data; in the writing of the manuscript; or in the decision to publish the results.

References

1. Poots, S.G.; Biesmeijer, J.C.; Kremen, C.; Neumann, P.; Schweiger, O.; Kunin, W.E. Global pollinator declines: Trends, impacts and drivers. *Trends Ecol. Evol.* **2010**, *25*, 345–353. [CrossRef] [PubMed]
2. Rollin, O.; Benelli, G.; Benvenuti, S.; Decourtye, A.; Wratten, S.D.; Canale, A.; Desneux, N. Weed-insect pollinator networks as bio-indicators of ecological sustainability in agriculture. A review. *Agron. Sustain. Dev.* **2016**, *36*, 1–22. [CrossRef]
3. Goulson, D.; Nicholls, E.; Botias, C.; Rotheray, E.L. Bee declines driven by combined stress from parasites, pesticides, and lack of flowers. *Science* **2015**, *347*, 1255957. [CrossRef] [PubMed]
4. Kluser, S.; Neumann, P.; Chauzat, M.P.; Pettis, J.S. UNEP Emerging Issues: Global Honey Bee Colony Disorder and other Threats to Insect Pollinators. United Nations Environment Programme 2010. Available online: https://europa.eu/capacity4dev/unep/document/global-bee-colony-disorders-and-other-threats-insect-pollinators (accessed on 30 October 2019).
5. Smith, K.M.; Loh, E.H.; Rostal, M.K.; Zambrana-Torrelio, C.M.; Mndiola, L.; Daszak, P. Pathogens, pests, and economics: Drivers of honey bee colony declines and losses. *EcoHealth* **2014**, *10*, 434–445. [CrossRef] [PubMed]
6. Ramsey, S.D.; Ochoa, R.; Bauchan, G.; Gulbronson, C.; Mowery, J.; Cohen, A.; Lim, D.; Joklik, J.; Cicero, J.M.; Ellis, J.D.; et al. *Varroa destructor* feeds primarily on honey bee fat body tissue not haemolymph. *Proc. Natl. Acad. Sci. USA* **2019**, *116*, 1792–1801. [CrossRef]
7. Ryabov, E.V.; Wood, G.R.; Fannon, J.M.; Moore, J.D.; Bull, J.C.; Chandler, D.; Mead, A.; Burroughs, N.; Evans, D.J. A Virulent Strain of Deformed Wing Virus (DWV) of Honeybees (*Apis mellifera*) Prevails after *Varroa destructor*-Mediated, or *in Vitro*, Transmission. *PLoS Pathog.* **2014**, *10*, e1004230. [CrossRef] [PubMed]
8. Nazzi, F.; Brown, S.P.; Annoscia, D.; Del Piccolo, F.; Di Prisco, G.; Varricchio, P.; Della Vedova, G.; Cattonaro, F.; Caprio, E.; Pennacchio, F. Synergistic parasite-pathogen interactions mediated by host immunity can drive the collapse of honey bee colonies. *PLoS Pathog.* **2012**, *8*, e1002735. [CrossRef]
9. Nazzi, F.; Le Conte, Y. Ecology of *Varroa destructor*, the Major Ectoparasite of the Western Honey Bee, *Apis mellifera*. *Annu. Rev. Entomol.* **2016**, *61*, 417–432. [CrossRef]
10. Yang, X.; Cox-Foster, D.L. Impact of an ectoparasite on the immunity and pathology of an invertebrate: Evidence for host immunosuppression and viral amplification. *Proc. Natl. Acad. Sci. USA* **2005**, *102*, 7470–7475. [CrossRef]
11. Gilioli, G.; Simonetto, A.; Hatina, F.; Sperandio, G. Multi-dimensional modelling tools supporting decision-making for the beekeeping sector. *IFAC PapersOnLine* **2018**, *51*, 144–149. [CrossRef]

12. Rortais, A.; Arnold, G.; Dorne, J.-L.; More, S.J.; Sperandio, G.; Streissl, F.; Szentes, C.; Verdonck, F. Risk assessment of pesticides and other stressors in bees: Principles, data gaps and perspectives from the European Food Safety Authority. *Sci. Total Environ.* **2017**, *587*, 524–537. [CrossRef] [PubMed]

13. Anonymous. *National Beekeeping Program for Period 2017–2019*; Ministry of Agriculture: Zagreb, Croatia, 2016.

14. Anonymous. *National Beekeeping Program for Period 2020–2022*; Ministry of Agriculture: Zagreb, Croatia, 2019.

15. Anonymous. *Program for Varroosis Control and Eradication*; Ministry of Agriculture: Croatia, 2017. Available online: http://www.veterinarstvo.hr/default.aspx?id=140 (accessed on 2 November 2019).

16. Anonymous. *Croatian National Regulation on Animal Protection Measures against Infectious and Parasitic Diseases and on Related Financing for 2019*; Ministry of Agriculture: Croatia, 2019. N.N. 5/2019. Available online: https://narodne-novine.nn.hr/clanci/sluzbeni/2019_01_5_92.html (accessed on 2 November 2019).

17. Rosenkranz, P.; Aumeier, P.; Ziegelmann, B. Biology and control of *Varroa destructor*. *J. Invertebr. Pathol.* **2010**, *103*, S96–S119. [CrossRef] [PubMed]

18. Frey, E.; Rosenkranz, P. Autumn invasion rates of *Varroa destructor* (Mesostigmata: Varroidae) into honey bee (Hymenoptera: Apidae) colonies and the resulting increase in mite populations. *J. Econ. Entomol.* **2014**, *107*, 508–515. [CrossRef] [PubMed]

19. Seeley, T.D.; Smith, M.L. Crowding honeybee colonies in apiaries can increase their vulnerability zo the deadly ectoparasite *Varroa destructor*. *Apidologie* **2015**, *46*, 716–727. [CrossRef]

20. Floris, I.; Satta, A.; Garau, V.L.; Melis, M.; Cabras, P.; Aloul, N. Effectiveness, persistence, and residue of amitraz plastic strips in the apiary control of *Varroa destructor*. *Apidologie* **2001**, *32*, 577–585. [CrossRef]

21. Sammataro, D.; Untalan, P.; Guerrero, F.; Finley, J. The resistance of varroa mites (Acari: Varroidae) to acaricides and the presence of esterase. *Int. J. Acarol.* **2005**, *31*, 67–74. [CrossRef]

22. Maggi, M.D.; Ruffinengo, S.R.; Negri, P.; Eguaras, M.J. Resistance phenomena to amitraz from populations of the ectoparasitic mite *Varroa destructor* of Argentina. *J. Parasitol. Res.* **2010**, *107*, 1189–1192. [CrossRef]

23. Howis, M.; Nowakowski, P. *Varroa destructor* removal efficiency using Beevital Clean preparation. *J. Apic. Sci.* **2009**, *53*, 15–20.

24. Rodriguez-Dehaibes, S.R.; Parido Sedas, V.T.; Luna-Olivars, G.; Villanuva-Jimenez, J.A. Two commercial formulations of natural compounds for *Varroa destructor* (Acari: Varroidae) control on Africanized bees under tropical climatic conditions. *J. Apic. Res.* **2017**, *56*, 58–62. [CrossRef]

25. Tlak Gajger, I.; Sušec, P. Efficacy of varroacidal food additive appliance during summer treatment of honeybee colonies (*Apis mellifera*). *Vet. Arhiv* **2019**, *89*, 87–96. [CrossRef]

26. Tlak Gajger, I.; Tomljanović, Z.; Stanisavljević, L. An environmentally friendly approach to the control of *Varooa destructor* mite and *Nosema ceranae* disease in Carnolian honeybee (*Apis mellifera carnica*) colonies. *Arch. Biol. Sci.* **2013**, *65*, 1585–1592. [CrossRef]

27. Anonymous. Guideline on Veterinary Medicinal Products Controlling *Varroa Destructor* Parasitosis in Bees. EMA— European Medicines Agency, EMA/CVMP/EWP/459883/2008, Committee for Medicinal Products for Veterinary Use (CVMP). Available online: http://www.ema.europaeu/docs/en_GB/document_library/Scientific_guideline/2010/11/WC500099137.pdf (accessed on 2 November 2019).

28. Delaplane, K.S.; Van Der Steen, J.; Guzman-Novoa, E. Standard methods for estimating strength parameters of *Apis mellifera* colonies. *J. Apic. Res.* **2013**, *52*, 1–12. [CrossRef]

29. Tlak Gajger, I.; Tomljanovic, Z.; Petrinec, Z. Monitoring health status of Croatian honey bee colonies and possible reasons for winter losses. *J. Apic. Res.* **2010**, *49*, 107–108. [CrossRef]

30. Mutinelli, F. Veterinary medicinal products to control *Varroa destructor* in honey bee colonies (*Apis mellifera*) and related EU legislation—An update. *J. Apic. Res.* **2016**, *55*, 78–88. [CrossRef]

31. Žorat, T. *The Efficiency of Acaricides used in Summer Treatment of Honeybee Colonies against Varroosis. Graduating Thesis*; University of Zagreb Faculty of Veterinary Medicine: Zagreb, Croatia, 10 October 2015.

32. Van Dooremalen, C.; Gerritsen, L.; Cornelissen, B.; van der Steen, J.J.M.; van Langevelde, F.; Blacquière, T. Winter Survival of Individual Honey Bees and Honey Bee Colonies Depends on Level of *Varroa destructor* Infestation. *PLoS ONE* **2012**, *7*, e36285. [CrossRef] [PubMed]

33. Gregorc, A.; Smodis Škerl, I.M. Combating *Varroa destructor* in Honeybee Colonies Using Flumethrin or Fluvalinate. *Acta Vet. Brno* **2007**, *76*, 309–314. [CrossRef]

34. Gregorc, A.; Planinc, I. Use of Thymol formulations, Amitraz, and Oxalic Acid for the control of the varroa mite in honey bee (*Apis mellifera carnica*) colonies. *J. Apic. Sci.* **2012**, *56*, 61–69. [CrossRef]

35. Blacquiere, T.; Altreuther, G.; Krieger, K.J. Evaluation of the Efficacy and Safety of Flumethrin 275 mg Bee-hive Strips (PolyVar Yellow®) against *Varroa destructor* in Naturally Infested Honey Bee Colonies in a Controlled Study. *Parasitol. Res.* **2017**, *116*, 109–122. [CrossRef]
36. Moro, A.; Mutinelli, F. Field evaluation of Maqs® and Api-Bioxal® for late summer control of Varroa mite infestation in Northeastern Italy. *J. Apic. Res.* **2018**, 53–61. [CrossRef]
37. Wilkinson, D.; Smith, G.C. A model of the mite parasite, *Varroa destructor*, on honeybees (*Apis mellifera*) to investigate parameters important to mite population growth. *Ecol. Model.* **2002**, *148*, 263–275. [CrossRef]

Article

Population Growth and Insecticide Residues of Honey Bees in Tropical Agricultural Landscapes

Damayanti Buchori [1,2], **Akhmad Rizali** [3,*], **Windra Priawandiputra** [4], **Dewi Sartiami** [1] **and Midzon Johannis** [5]

[1]   Department of Plant Protection, Faculty of Agriculture, Bogor Agricultural University (IPB University), Bogor 16680, Indonesia; damayanti@apps.ipb.ac.id (D.B.); dsartiami@apps.ipb.ac.id (D.S.)
[2]   Center for Transdisciplinary and Sustainability Sciences, Bogor Agricultural University (IPB University), Bogor 16129, Indonesia
[3]   Department of Plant Pests and Diseases, Faculty of Agriculture, University of Brawijaya, Malang 65145, Indonesia
[4]   Department of Biology, Faculty Mathematics and Natural Sciences, Bogor Agricultural University (IPB University), Bogor 16680, Indonesia; priawandiputra@apps.ipb.ac.id
[5]   PT. Syngenta Indonesia, Jakarta 12560, Indonesia; midzon.johannis@syngenta.com
*   Correspondence: arizali@ub.ac.id; Tel./Fax: +62-341-575843

Received: 14 November 2019; Accepted: 16 December 2019; Published: 18 December 2019

Abstract: Global decline of pollinators, especially bees, has been documented in many countries. Several causes such as land-use change and agricultural intensification are reported to be the main drivers of the decline. The objective of this study was to investigate the effect of land use on honey bee and stingless bee populations. Research was conducted in Bogor and Malang to compare between two different geographical areas. Managed bees such as honey bees (*Apis cerana* and *A. mellifera*) and stingless bees (*Tetragonula laeviceps*) were investigated to examine the effect of agricultural intensification. Field experiments were conducted by placing beehives in selected habitats (i.e., beekeeper gardens, forests areas, and agriculture areas). Population growth and neonicotinoid residue analysis of bees in different hive locations were measured to study the effect of habitat type. Population growth of bees represents the forager abundance and colony weight. Based on the analysis, we found that habitat type affected forager abundance and colony weight of honey bees ($p < 0.05$), although the patterns were different between species, region, as well as season. Forests could support the stingless bee colony better than agriculture and home garden habitats. Insecticide (neonicotinoid) was barely recorded in both honey bees and stingless bees.

Keywords: *Apis cerana*; *Apis mellifera*; agriculture; forests; home garden; neonicotinoid; *Tetragonula laeviceps*

---

1. Introduction

In the agroecosystem, pollinators are a pivotal component of biodiversity that provide an important ecosystem service through crop pollination [1] and increasing fruit set [2]. Pollinators also can be used as indicators of ecosystem health because of their sensitiveness to environmental stressors [3], for instance, the negative impacts of pesticide application [4]. There is growing concern relating to declines found in pollinators around the world [5]. In Europe and the US, a decline in wild bee species richness has been recorded, where the declining trends are in the abundance of honey bees (*Apis mellifera*) and a small number of wild pollinators [6]. Although high diversity of bees is found in the tropics (e.g., [7,8]), there is a lack of information about this phenomenon. Therefore, investigation needs to be undertaken into the scale, magnitude, and causes of the decline and the effects on pollination services.

Global declines in honey bees and wild bees have been associated to habitat loss and fragmentation, pesticide application, pathogens, invasive species, and climate change [9,10]. The potential threat of insecticides, such as neonicotinoids, for honey bees and wild bees has been reported, although it is still in debate [11]. Neonicotinoids have negative impacts such as increasing the mortality of honey bees by impairing their homing ability [12] and reducing the reproductive success of bumble bees and solitary bees [13], although other studies have reported no effects [13]. There is limited information from comprehensive studies on the impact of neonicotinoids toward long-term survival of honey bee colonies [14]. Landscape-scale experiments in different geographical regions are needed to investigate the impacts of neonicotinoids on bees [13,14].

The research was conducted in various land-use types both in Bogor (West Java) and Malang (East Java), Indonesia. Bogor has unique agricultural characteristics, as it is surrounded by mountain areas, and has a seminatural habitat dominated by agricultural fields cultivated with rotations of crop plants, which are mainly rice and vegetables [15]. Similar to Bogor, Malang is also surrounded by mountain areas and consists of tropical rainforests as well as cultivated and settlement areas. Agriculture is the primary land-use on the island of Java, so its management has profound consequences for the environment and for biodiversity. Agricultural intensification, especially pesticide application, is commonly used as a consequence of the green revolution [16] and has a negative effect on biodiversity, especially pollinators [17].

The objectives of this research were to investigate land-use effects and the indirect impact of insecticide application in agricultural areas on insect pollinators, particularly bees. Most evidence of the impact of pesticides on pollinators, especially honey bees, has come from laboratory-based toxicity tests. Negative effects of insecticide have been reported (e.g., [18]), but field research is still needed to understand how laboratory-derived toxicity levels effect pollinator communities in the agroecosystem, although some field- and semi-field-based studies have been conducted [19].

2. Materials and Methods

2.1. Research Site and Experiment Plot Selection

Research sites were located in West Java (Bogor and its surrounding area) and East Java (Malang and its surrounding area) to compare between two different geographical areas (Figure 1). Both areas are reported as honey producers in Indonesia [20]. Beekeepers in Bogor were characterized by breeding in a lower elevation (<300 m asl), while in Malang it was in a higher elevation (>500 m asl) (https://en.climate-data.org). To conduct the experiment, we selected two different habitat types, agriculture and forest, both in Bogor and Malang. As a comparison, we also observed the hives in selected beekeeper gardens (home garden) (Table 1). We used three different species of bees during the experiment that were commonly breed by beekeepers: *A. mellifera* and *A. cerana* in Malang and *A. cerana* and *T. laeviceps* in Bogor. In each habitat, we selected three experiment plots in different locations for replication, and the minimum distance between plots was 2 km (Figure 1).

Table 1. Satellite images and description of habitat types for research experiments in Bogor and Malang. Satellite images were derived from Google Maps, accessed year 2019.

Area	Agriculture	Forest	Home Garden
Bogor			
Description	Dominated by rice, maize, and cucumber with frequent pesticide application	Tree plants, dominated by *Paraserienthes falcataria* or *Hevea braziliensis*	Habitat surrounding housing area, dominated by fruit trees and flowering plants

Table 1. *Cont.*

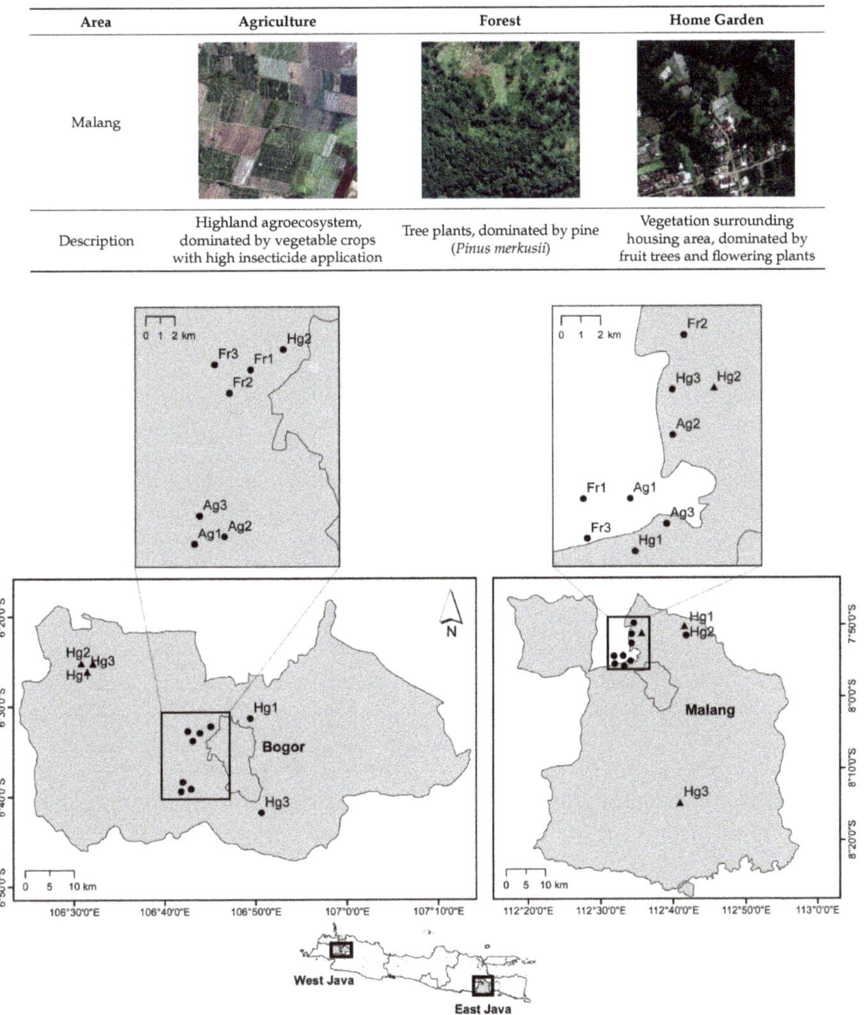

Area	Agriculture	Forest	Home Garden
Malang			
Description	Highland agroecosystem, dominated by vegetable crops with high insecticide application	Tree plants, dominated by pine (*Pinus merkusii*)	Vegetation surrounding housing area, dominated by fruit trees and flowering plants

Figure 1. Map of study sites for research experiments in different habitat types (Ag: agriculture, Fr: forest, Hg: home garden (beekeepers place) in Bogor (West Java) and Malang (East Java)). The numbers after the habitat code indicates plot number. Home garden with triangle symbol indicates beekeepers of *Apis cerana*.

We placed three beehives in each plot for observation. The sizes of beehives were different depending on the beekeeper's practice in rearing bees. In Malang, the hive size of *A. cerana* was 40 × 25 × 25 cm with four combs and *A. mellifera* was 50 × 40 × 25 cm with four combs. While in Bogor, the hive size of *A. cerana* was 35 × 30 × 25 cm with six combs and *T. laeviceps* was 30 × 10 ×10 cm without combs.

2.2. Observation of Bees in the Hives and Residue Analysis

To study the effect of habitat type on bees, we observed the hives on each experimental plot, measured the population growth of bees, and performed a residue analysis. Population growth of bees was measured by counting the forager abundance (foraging activity) and colony weight (weighing full hive). The method of foraging activity monitoring was based on [21] by counting the foragers departing or returning to the colony for thirty minutes per hive. In each plot, the observation was conducted from 7 a.m. until 11 a.m. Weighing full hives was conducted to calculate the colony weight that included the summed weight of the box, combs with food stores, and the bees [22]. Both forager abundance and colony weight were observed every two weeks during two months. Observations were conducted in different seasons (i.e., rainy and dry seasons). Observations in the dry season were done from March to May 2019, while for the rainy season, observations were from July to September 2019.

In addition, insecticide residue analysis was conducted by collecting honey and bee (foragers) samples in three habitat types. We collected 5 mg of bees and 5 mg of honey per plot and initially froze it before being analyzed using the QuEChERS protocol [23]. Insecticide residue analysis was conducted in the medical laboratory of Jakarta (https://labkesda.jakarta.go.id) and was focused on imidacloprid content as the representation of neonicotinoid insecticide.

2.3. Data Analysis

The difference of forager abundance and colony weight of bees between habitat types was analyzed using a Kruskal–Wallis test. If we found significant differences, a post-hoc test was done using Fisher's least-significant difference with $\alpha = 0.05$. To analyze the relationship between forager abundance as well as colony weight of bees and observation time, we analyzed using fitting median-based linear models based on the Theil–Sen single median. All analyses were performed using R statistical software [24] and package "mblm" for fitting median-based linear models [25].

3. Results

3.1. Effect of Different Habitat Types and Season on Honey Bees

Based on the analysis, we found that habitat type and season affected the forager abundance and colony weight of bees ($p < 0.05$), although the patterns were different between species and region. In general, habitat types showed to be the most important factor that affected both forager abundance and colony weight of bees compared to season and observation time (Table 2). Effect of habitat type on forager abundance and colony weight of *A. cerana* differed between Malang and Bogor. In Malang, the difference of habitat type significantly influenced the forager abundance and colony weight of *A. cerana*, which were found higher in home gardens and forests compared to the agricultural areas (Figures 2a and 3a). In Bogor, forager abundance of *A. cerana* was not significantly different among habitat type (Figure 2b), while the colony weight of *A. cerana* in forests was significantly higher than in home gardens (Figure 3b). The same pattern was observed with *A. cerana* in Malang for forager abundance of *A. mellifera* in Malang and *T. laeviceps* in Bogor, which were also found higher in home gardens and forests compared to agricultural areas (Figure 2c,d and Figure 3c,d).

Table 2. The results of Kruskal–Wallis tests of forager abundance and colony weight of bees in different seasons, habitat types, and observation times.

Variable	Bee	Region	Season (df = 1)		Habitat Type (df = 2)		Observation Time (df = 3)	
			χ^2	*p*-Value	χ^2	*p*-Value	χ^2	*p*-Value
Forager abundance	*Apis cerana*	Malang	0.960	0.327	22.448	<0.001	4.430	0.219
		Bogor	0.006	0.937	3.804	0.149	0.239	0.971
	Apis mellifera	Malang	0.323	0.570	18.543	<0.001	0.106	0.991
	Tetragonula laeviceps	Bogor	0.051	0.822	17.106	<0.001	1.159	0.763
Colony weight	*A. cerana*	Malang	2.144	0.143	8.737	0.013	0.581	0.901
		Bogor	4.202	0.040	4.425	0.109	3.383	0.336
	A. mellifera	Malang	18.365	<0.001	9.092	0.011	0.056	0.997
	T. laeviceps	Bogor	0.011	0.915	45.094	<0.001	1.615	0.656

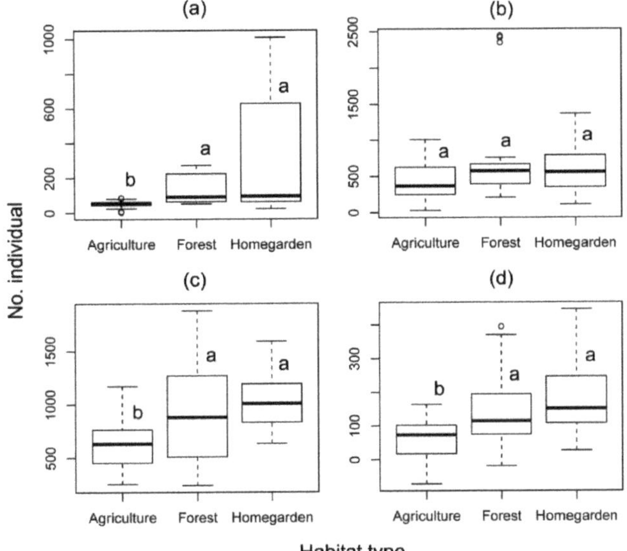

Figure 2. Forager abundance of honey bees and stingless bees in different habitats. (**a**) *A. cerana* in Malang, (**b**) *A. cerana* in Bogor, (**c**) *A. mellifera* in Malang, and (**d**) *T. laeviceps* in Bogor. Boxes with different letters are significantly different at $p < 0.05$ according to Fisher's least-significant difference.

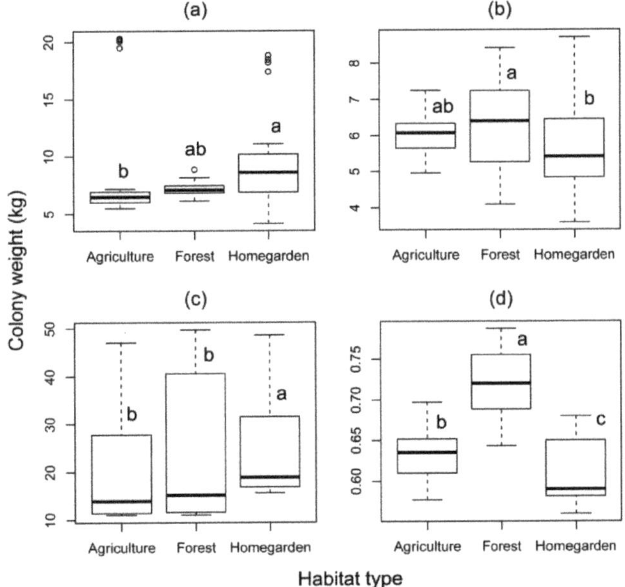

Figure 3. Colony weight of honey bees and stingless bees in different habitats. (**a**) *A. cerana* in Malang, (**b**) *A. cerana* in Bogor, (**c**) *A. mellifera* in Malang, and (**d**) *T. laeviceps* in Bogor. Box with different letters are significantly different at $p < 0.05$ according to Fisher's least significant difference.

Based on non-parametric regression, we found that the forager abundance of *A. cerana* in Malang was prone to decrease with increasing observation time ($p = 0.001$) (Table 3). In addition, the colony weights of *A. cerana* and *T. laeviceps* in Bogor as well as *A. mellifera* in Malang tended to increase with increasing observation time ($p < 0.05$).

Table 3. Relationship between forager abundance and colony weight of bees and observation time (day) based on non-parametric regression.

Parameter	*A. cerana*				*A. mellifera*		*T. laeviceps*	
	Malang		Bogor		Malang		Bogor	
	Estimate	*p*-Value	Estimate	*p*-Value	Estimate	*p*-Value	Estimate	*p*-Value
Forager abundance								
(Intercept)	89.000	<0.001	530.500	<0.001	783.750	<0.001	105.500	<0.001
Observation	−0.639	0.001	−0.342	0.567	1.700	0.637	0.339	0.160
Colony weight								
(Intercept)	7.095	<0.001	5.738	<0.001	15.915	<0.001	0.646	<0.001
Observation	−0.005	0.167	0.013	0.001	0.014	0.966	0.000	0.033

3.2. Detection Results of Insecticide Residue in Honey and Body of Honey Bees

Based on residue analysis, we barely detected insecticide residue (imidacloprid content) both in the honey and the body of bees (Table 4). Two of 18 honey samples (11.11%) contained a small amount of insecticide residue (<5 µg/kg). Insecticide residues were also detected in three bee samples (16.66%). Surprisingly, we did not detect the insecticide residue in the agriculture habitat, yet it was detected in forest and home garden habitats.

Table 4. Insecticide residue (imidacloprid content) detected from honey and bee bodies.

Species	Product	Habitat	Residue (µg/kg)	
			Bogor	Malang
A. mellifera	Honey	Agriculture	-	-
		Forest	4.4	-
		Home garden	-	-
	Bee body	Agriculture	-	-
		Forest	-	-
		Home garden	-	-
A. cerana	Honey	Agriculture	-	-
		Forest	-	0.5
		Home garden	-	-
	Bee body	Agriculture	-	-
		Forest	3.1	-
		Home garden	11.2	-
T. laeviceps	Honey	Agriculture	-	-
		Forest	-	-
		Home garden	-	-
	Bee body	Agriculture	-	-
		Forest	-	-
		Home garden	-	2.9

4. Discussion

Our research provided the evidence that habitat type significantly affects both forager abundance and colony weight of honey bees as well as stingless bees in Indonesia. The lowest forager abundance and colony weight was shown in agricultural areas, which indicated that agricultural areas had a negative effect on bees. Research by [26] showed that bee species have distinct preferences for different plant communities, and their abundance is related to the abundance of their host plants. Agricultural areas that are dominated by certain crop plants might affect the fitness and population growth of bees. This is due to plant diversity and is a key driver of bee fitness. Bees were found to be fitter and their populations grew faster in more florally diverse environments because of a continuous supply of food resources [27].

Landscape diversity also influences the growth and reproduction of honey bees, besides the availability of pollen in agricultural landscapes. For instance, *A. mellifera* compensated for lower landscape diversity by increasing their pollen foraging range in order to maintain pollen amounts and diversity [28]. This indicates the importance of agri-environmental schemes to support pollinators and not just the plant diversity and pollen availability. In Malang, *A. mellifera* is handled by beekeepers following "migratory management", which causes bees undue stress. In order to ease access to food source, beekeepers move the hives to flowering areas. However, this management may affect the population growth of bees. Research by [29] showed that the lifespan of migratory adult bees tends to decrease compared to stationary bees.

In this research, we only focused on environmental stressors (e.g., habitat condition) and did not investigate other potential drivers that affect population growth of bees, such as pests and pathogens, and genetic diversity as well as vitality of bees [9]. However, habitat type did not guarantee that bees were unhampered from insecticide. This might be related to the foraging range of bees as well as the food source, which is not only pollen but also honeydew. Honey bee presence was positively affected by the presence of honeydew and source of insecticide residue that affected honey quality [30]. Neonicotinoid (imidacloprid), which was found in our study, was also detected in 11% samples of honey from apiaries located in Poland [31]. However, the death of honey bees from Bologna were reported from different active ingredients of insecticide [32]. Thus, in this research, neonicotinoid was detected in the body of bees and honey, although it was in a small amount of residue.

5. Conclusions

Our research results showed the factor that affected forager abundance and colony weight of bees the most was habitat type. The agricultural habitat had lower bee forager abundance and colony weight compared to forest and home garden habitats. This indicates that the hypothesis from the beekeepers' perspective is accepted in relation to the negative effect of agriculture (especially pesticide application) on their honey production. However, our experiment revealed that habitat type, especially forest and home garden, did not guarantee that honey bees were unhampered from pesticide.

Author Contributions: Conceptualization, D.B., A.R., W.P., and M.J.; methodology, D.B., A.R., and W.P.; formal analysis, A.R. and W.P.; writing—original draft preparation, D.B., A.R., and W.P.; writing—review and editing, D.B., A.R., W.P., D.S., and M.J. All authors have read and agreed to the published version of the manuscript.

Funding: This research was funded by PT. Syngenta Indonesia in collaboration with the Entomological Society of Indonesia.

Acknowledgments: We thank Bella, Vina, Widia, Irline, Sefin, and Kartika who help during field observation.

Conflicts of Interest: Midzon Johannis is an employee of PT. Syngenta Indonesia.

References

1. Klein, A.M.; Vaissiere, B.E.; Cane, J.H.; Steffan-Dewenter, I.; Cunningham, S.A.; Kremen, C.; Tscharntke, T. Importance of pollinators in changing landscapes for world crops. *Proc. R. Soc. Lond. B* **2007**, *274*, 303–313. [CrossRef]
2. Klein, A.M.; Steffan-Dewenter, I.; Tscharntke, T. Fruitset of high land coffee increases with the diversity of pollinating bees. *Proc. R. Soc. Lond. B* **2003**, *270*, 955–961. [CrossRef] [PubMed]
3. Klein, S.; Cabirol, A.; Devaud, J.M.; Barron, A.B.; Lihoreau, M. Why bees are so vulnerable to environmental stressors? *Trends Ecol. Evol.* **2017**, *32*, 268–278. [CrossRef] [PubMed]
4. Kevan, P.G. Pollinators as bioindicators of the state of the environment: Species, activity and diversity. *Agric. Ecosyst. Environ.* **1999**, *74*, 373–393. [CrossRef]
5. Potts, S.G.; Imperatriz-Fonseca, V.; Ngo, H.T. *The Assessment Report on Pollinators, Pollination and Food Production of the Intergovernmental Science-Policy Platform on Biodiversity and Ecosystem Services*; Intergovernmental Science-Policy Platform on Biodiversity and Ecosystem Services (IPBES): Bonn, Germany, 2017.

6. Biesmeijer, J.C.; Roberts, S.P.M.; Reemer, M.; Ohlemuller, R.; Edwards, M.; Peeters, T.; Schaffers, A.P.; Potts, S.G.; Kleukers, R.; Thomas, C.D. Parallel declines in pollinators and insect pollinated plants in Britain and the Netherlands. *Science* **2006**, *313*, 351–354. [CrossRef]

7. Liow, L.H.; Sodhi, N.S.; Elmqvist, T. Bee diversity along a disturbance gradient in tropical lowland forests of south-east Asia. *J. Appl. Ecol.* **2001**, *38*, 180–192. [CrossRef]

8. Basu, P.; Parui, A.K.; Chatterjee, S.; Dutta, A.; Chakraborty, P.; Roberts, S.; Smith, B. Scale dependent drivers of wild bee diversity in tropical heterogeneous agricultural landscapes. *Ecol. Evol.* **2016**, *6*, 6983–6992. [CrossRef]

9. Potts, S.G.; Biesmeijer, J.C.; Kremen, C.; Neumann, P.; Schweiger, O.; Kunin, W.E. Global pollinator declines: Trends, impacts and drivers. *Trends Ecol. Evol.* **2010**, *25*, 345–353. [CrossRef]

10. Vanbergen, A.J. The insect pollinators initiative threats to an ecosystem service: Pressures on pollinators. *Front. Ecol. Environ.* **2013**, *11*, 251–259. [CrossRef]

11. Woodcock, B.A.; Bullock, J.M.; Shore, R.F.; Heard, M.S.; Pereira, M.G.; Redhead, J.; Ridding, L.; Dean, H.; Sleep, D.; Henrys, P.; et al. Country-specific effects of neonicotinoid pesticides on honey bees and wild bees. *Science* **2017**, *356*, 1393–1395. [CrossRef]

12. Henry, M.; Béguin, M.; Requier, F.; Rollin, O.; Odoux, J.-F.; Aupinel, P.; Aptel, J.; Tchamitchian, S.; Decourtye, A. A common pesticide decreases foraging success and survival in honey bees. *Science* **2012**, *336*, 348–350. [CrossRef] [PubMed]

13. Rundlof, M.; Andersson, G.K.; Bommarco, R.; Fries, I.; Hederstrom, V.; Herbertsson, L.; Jonsson, O.; Klatt, B.K.; Pedersen, T.R.; Yourstone, J.; et al. Seed coating with a neonicotinoid insecticide negatively affects wild bees. *Nature* **2015**, *521*, 77–80. [CrossRef] [PubMed]

14. Cutler, G.C.; Scott-Dupree, C.D.; Sultan, M.; McFarlane, A.D.; Brewer, L. A large-scale field study examining effects of exposure to clothianidin seed-treated canola on honey bee colony health, development, and overwintering success. *PeerJ* **2014**, *2*, e652. [CrossRef] [PubMed]

15. Widiatmaka, W.; Ambarwulan, W. Sudarsono Spatial multi-criteria decision making for delineating agricultural land in Jakarta metropolitan area's hinterland: Case study of Bogor regency, West Java. *AGRIVITA J. Agric. Sci.* **2016**, *38*, 105–115.

16. Hansen, G.E. Indonesia's Green Revolution: The Abandonment of a Non-Market Strategy toward Change. *Asian Surv.* **1972**, *12*, 932–946. [CrossRef]

17. Mariyono, J.; Kompas, T.; Grafton, R.Q. Shifting from Green Revolution to environmentally sound policies: technological change in Indonesian rice agriculture. *J. Asia Pac. Econ.* **2010**, *15*, 128–147. [CrossRef]

18. El Hassani, A.K.; Dacher, M.; Gauthier, M.; Armengaud, C. Effects of sublethal doses of fipronil on the behavior of the honeybee (*Apis mellifera*). *Pharmacol. Biochem. Behav.* **2005**, *82*, 30–39. [CrossRef]

19. Koch, H.; Weißer, P. Exposure of honey bees during pesticide application under field conditions. *Apidologie* **1997**, *28*, 439–447. [CrossRef]

20. BPS-Statistics Indonesia. *Statistics of Forestry Production 2017*; BPS-Statistics Indonesia: Jakarta, Indonesia, 2017.

21. Abou-Shaara, H.F.; Al-Ghamdi, A.A.; Mohamed, A.A. Honey bee colonies performance enhance by newly modified beehives. *J. Apic. Sci.* **2013**, *57*, 45–57. [CrossRef]

22. Human, H.; Brodschneider, R.; Dietemann, V.; Dively, G.; Ellis, J.D.; Forsgren, E.; Fries, I.; Hatjina, F.; Hu, F.-L.; Jaffé, R.; et al. Miscellaneous standard methods for *Apis mellifera* research. *J. Apic. Res.* **2013**, *52*, 1–53. [CrossRef]

23. Calatayud-Vernich, P.; Calatayud, F.; Simo, E.; Pico, Y. Efficiency of QuEChERS approach for determining 52 pesticide residues in honey and honey bees. *MethodsX* **2016**, *3*, 452–458. [CrossRef] [PubMed]

24. R Core Team. *R: A Language and Environment for Statistical Computing*; R Foundation for Statistical Computing: Vienna, Austria, 2019.

25. Komsta, L. *mblm: Median-Based Linear Models*, R package version 0.12.1; 2019. Available online: https://CRAN.R-project.org/package=mblm (accessed on 14 November 2019).

26. Wu, P.; Axmacher, J.C.; Song, X.; Zhang, X.; Xu, H.; Chen, C.; Yu, Z.; Liu, Y. Effects of plant diversity, vegetation composition, and habitat type on different functional trait groups of wild bees in rural Beijing. *J. Insect Sci.* **2018**, *18*. [CrossRef] [PubMed]

27. Kaluza, B.F.; Wallace, H.M.; Heard, T.A.; Minden, V.; Klein, A.; Leonhardt, S.D. Social bees are fitter in more biodiverse environments. *Sci. Rep.* **2018**, *8*, 12353. [CrossRef] [PubMed]

28. Danner, N.; Keller, A.; Hartel, S.; Steffan-Dewenter, I. Honey bee foraging ecology: Season but not landscape diversity shapes the amount and diversity of collected pollen. *PLoS ONE* **2017**, *12*, e0183716. [CrossRef]
29. Simone-Finstrom, M.; Li-Byarlay, H.; Huang, M.H.; Strand, M.K.; Rueppell, O.; Tarpy, D.R. Migratory management and environmental conditions affect lifespan and oxidative stress in honey bees. *Sci. Rep.* **2016**, *6*, 32023. [CrossRef]
30. Robertson, L.M.; Edlin, J.S.; Edwards, J.D. Investigating the importance of altitude and weather conditions for the production of toxic honey in New Zealand. *N. Z. J. Crop Hortic. Sci.* **2010**, *38*, 87–100. [CrossRef]
31. Bargańska, Ż.; Ślebioda, M.; Namieśnik, J. Pesticide residues levels in honey from apiaries located of Northern Poland. *Food Control* **2013**, *31*, 196–201. [CrossRef]
32. Porrini, C.; Sabatini, A.G.; Girotti, S.; Fini, F.; Monaco, L.; Celli, G.; Bortolotti, L.; Ghini, S. The death of honey bees and environmental pollution by pesticides: The honey bees as biological indicators. *Bull. Insectol.* **2003**, *56*, 147–152.

Review

Impact of Stressors on Honey Bees (*Apis mellifera*; Hymenoptera: Apidae): Some Guidance for Research Emerge from a Meta-Analysis

Tiphaine Havard [1,*], Marion Laurent [2] and Marie-Pierre Chauzat [2,3]

[1] Regulated Product Assessment Department, ANSES, 94700 Maisons-Alfort, France
[2] Unit of Honey Bee Pathology, ANSES, European Union and National Reference Laboratory for Honey Bee Health, Sophia Antipolis, 06560 Alpes-Maritimes, France; marion.laurent@anses.fr
[3] Strategy and Programs Department, 94700 Maisons-Alfort, France; marie-pierre.chauzat@anses.fr
* Correspondence: tiphaine.havard@anses.fr

Received: 14 November 2019; Accepted: 19 December 2019; Published: 20 December 2019

Abstract: Bees play an essential role in plant pollination and their decline is a threat to crop yields and biodiversity sustainability. The causes of their decline have not yet been fully identified, despite the numerous studies that have been carried out, especially on *Apis mellifera*. This meta-analysis was conducted to identify gaps in the current research and new potential directions for research. The aim of this analysis of 293 international scientific papers was to achieve an inventory of the studied populations, the stressors and the methods used to study their impact on *Apis mellifera*. It also aimed to investigate the stressors with the greatest impact on bees and explore whether the evidence for an impact varies according to the type of study or the scale of study. According to this analysis, it is important to identify the populations and the critical developmental stages most at risk, and to determine the differences in stress sensibility between subspecies. This meta-analysis also showed that studies on climate change or habitat fragmentation were lacking. Moreover, it highlighted that technical difficulties in the field and the buffer effect of the colony represent methodological and biological barriers that are still difficult to overcome. Mathematical modeling or radio frequency identification (RFID) chips represent promising ways to overcome current methodological difficulties.

Keywords: populations under study; biological effects; stress; experimental methods; techniques; honey bees

---

1. Introduction

Honey bees are important pollinators of most wild plants [1] and agricultural crops [2]. They are the most economically important group of pollinators worldwide [3] and are also crucial for maintaining biodiversity [4]. In particular, the economic contribution of the honey bee *Apis mellifera* to agriculture is estimated at USD 20 billion in the US and more than USD 200 billion worldwide [5].

Over the past decades, significant losses of wild and domestic bees have been reported in many parts of the world [6], threatening the ecosystem services they provide. Many hypotheses have been put forward to explain these losses, but the causes are not yet clearly identified [3]. So far, no single factor appears to act as the main driver of bee decline [7,8] and this phenomenon is now widely regarded as multifactorial [6–10]. Among the factors involved, biological and chemical agents are at the forefront. Indeed, bees are chronically exposed to pesticide cocktails, but also to many parasitic and infectious agents (PIAs), some of which are still emerging as they are disseminated by humans and international transport [9]. In addition, other stressors such as habitat loss, beekeeping practices, climate change or decreased abundance and diversity of floral resources are likely to contribute to the decline of bees, making them more sensitive to other stressors [9]. This multiplicity of stressors makes the diagnosis of bee decline all the more complicated given that bees are exposed simultaneously or successively to several stressors, resulting in many interactive and synergistic effects [11].

Although mass mortality is the most striking impact of stressors on bees, some non-lethal effects can also lead to important losses. For example, one hypothesis to explain recurrent weakening of honey bee colonies is that disoriented bees are no longer able to find their way home [12]. Thus, the effect of stressors on bee health is not limited to lethal effects but is also related to behavioral changes [13], impaired cognitive functions and sensory abilities [14,15], and physiological [16], molecular [17] and genetic changes [10]. At present, the effects of pesticides and other stressors on bee health are still poorly understood and are not evaluated by standard regulatory procedures for risk assessment [18]. The economic and ecological challenges represented by bee decline explain the current search for risk assessment procedures and methods to analyze the impact of stressors on bee health [19]. This meta-analysis was conducted to identify gaps in the current research and new potential directions for research.

The aims of this meta-analysis were (i) to carry out an inventory of the bee populations under study, the stressors studied and the methods used by the scientists, (ii) to investigate whether the stressors with the greatest impact on bees could be identified, and (iii) to explore whether the evidence for an impact varies according to the type of study or to the scale of study. This work follows a bibliographic study carried out in 2016 on the exposure of honey bees to plant protection products [20].

2. Materials and Methods

2.1. Identification of the Key Concepts and the Relevant Keywords

The population, exposure, outcomes (PEO) method was used to define the key concepts of the analysis. Three key concepts were then identified: the target population, the stressors studied and the methods used. Keywords were listed for each key concept after reading a subset of scientific papers related to the impact of stressors on *Apis mellifera*. A search was subsequently performed in Scopus and Cab Abstract databases with these keywords (see Figure S1 in the Supplementary Materials for details of the search string) and resulted in the selection of 3999 articles.

2.2. Literature Search

The target population included in the study comprised subspecies of the honey bee *Apis mellifera* with the exception of *A. m. scutellata*, *A. m. capensis* and Africanized bees. All epidemiological units (colony, adults, brood) and development stages (eggs, larvae, pupa) were included in the analysis. The papers included in this survey were published during the last ten years (from 2007 to 2017, last access: 6 March, 2017). The articles were available in full text and written in English. The primary search in the Scopus and Cab Abstract databases resulted in the selection of 3999 articles; 1187 duplicates were removed (Figure 1), then 717 articles were excluded because they dealt with *A. m. scutellata* or *A. m. capensis* (82 articles), with other organisms (73 articles), with the efficiency of veterinary treatments (117 articles), with the presence of pesticides in bee matrices (26 articles) or because they were off topic (419 articles).

A new search was performed on the remaining 2095 articles to better focus on the impact of stress on honey bees. To be included in the analysis, the title of the articles had to contain the following words: ("honey bee" or "mellifera") and ("impact" or "affect" or "effect" or "influence" or "toxicity" or "impair" or "induce"). Following this procedure, 386 articles were selected. Reviews and articles dealing with stressors considered as "anecdotal" (i.e., stressors to which bees are rarely exposed for example caffeine, nanoparticles; see details in Figure S2, Supplementary Materials) were excluded from the list. After this selection process, 293 articles were included in the analysis (see Table S1 in Supplementary Materials for the references). Although our paper selection was implemented thoroughly, we acknowledge that some references may have been omitted, however we believe that the number of these references is very small.

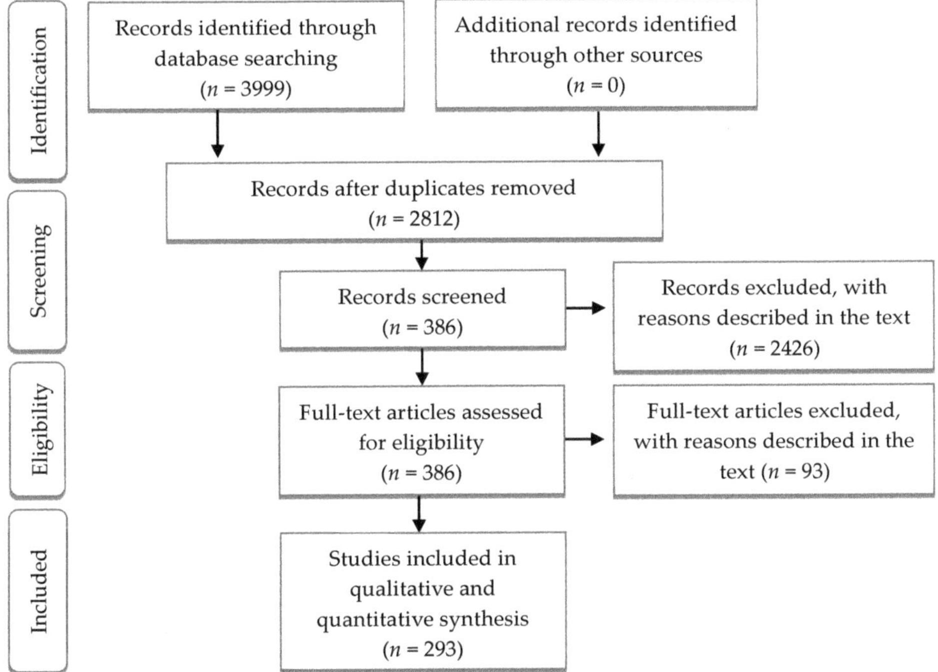

Figure 1. Preferred reporting items for systematic reviews and meta-analyses (PRISMA) flow diagram describing the meta-analysis of literature on the impact of stressors on honey bees' health—summary of how the systematic search was conducted and eligible studies were identified.

2.3. Data Extraction

Each publication ($n = 293$) was reviewed using a standard protocol. General information about the authors affiliation, country, the year of publication and the journal was recorded. Information on the subspecies, the bee life stages (larvae, pupae or adults, i.e., workers, queens, drones), and the stressors studied in the publication, the endpoint measurements and the methods used were stored in a database dedicated to the analysis. Whether several stressors were tested simultaneously or not was noted. We also recorded whether the impact of the stressor was evidenced or not. Finally, we included the scale of study in the analysis. The endpoint measurements were grouped into four classes: (i) colony scale (e.g., colony weight, colony reproduction, colony survival), (ii) individual scale (e.g., physiological or anatomical measures, learning, memory, behavior, mortality), (iii) cellular scale (e.g., cell death, spermatozoa viability), (iv) molecular scale (e.g., enzyme activity, protein concentration), and (v) genetic scale (e.g., genes expression).

2.4. Data Analysis

Flows between two or more variables are represented by Sankey diagrams, in which the width of the arrows is proportional to the magnitude of the flow. Diagrams were produced with the online tool SankeyMATIC.

Statistical analysis was conducted using Chi-square tests implemented with R.

3. Results and Discussion

3.1. General Information

The articles analyzed in this meta-analysis (*n* = 293) were published in 128 different journals, the three most frequent being *Plos One* (13%), *Apidologie* (10%), and the *Journal of Apicultural Research* (6%) (Supplementary Materials, Figure S3). The number of papers published per year increased relatively steadily (Figure 2) from 2010 (24 papers) to 2016 (54 papers). This could be a sign of the scientific community's growing interest in bee health and especially its willingness to investigate the mechanisms leading to abnormal bee mortality.

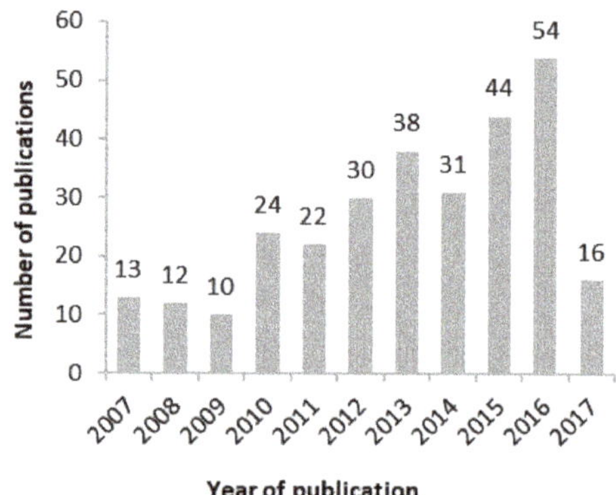

Figure 2. Number of publications (*n* = 293) related to the impact of stressors on *Apis mellifera* published per year between 2007 and 2017 and included in the study (last access to database: 6 March 2017).

3.2. What Kind of Populations were Studied in the Articles?

Worker bees were by far the most studied category of bees (67%, *n* = 230) and of these, foragers were widely represented (17% of the total population) (Figure 3a). Brood and colonies were studied in 13% of cases (*n* = 43), while drones (male bees) and queens were rarely studied (2% and 5% of the publications, respectively). Queens have been studied since 2011 whereas drones have only been included in papers since 2013 (see Figure S4 in the Supplementary Materials). Workers, and especially foragers, are the first to suffer from abnormally high mortality rates and seem to be the age class most exposed to stressors. Moreover, due to their number they constitute the greater part of the colony and probably represent the simplest biological material to study in hives. Drones and queens are key elements for colony survival due to their role in reproduction and recent studies have revealed that their reproductive capacities are altered by stressors [21–28]. In addition, nurses play a decisive role in larvae development due to the quality of the food they produce. However, bee nurses together with queens and drones are poorly studied when compared to bee workers.

More than half of the authors did not specify which *A. mellifera* subspecies was used in their experiments (Figure 3b), probably because there is numerous inter-subspecific cross-breeding in the field that makes identification difficult. However, stress sensitivity may differ between two subspecies [29,30]. Therefore, the subspecies is an important parameter to take into account and should be documented. When specified, the most studied subspecies were *A. m. carnica* (13%) and *A. m. ligustica* (9%) followed by Buckfast bees (5%) and *A. m. mellifera* (2%).

Therefore, it seems important to identify the populations most at risk or the critical developmental stages, to identify the differences in stress sensitivity between subspecies and potentially to define one or two indicative subspecies.

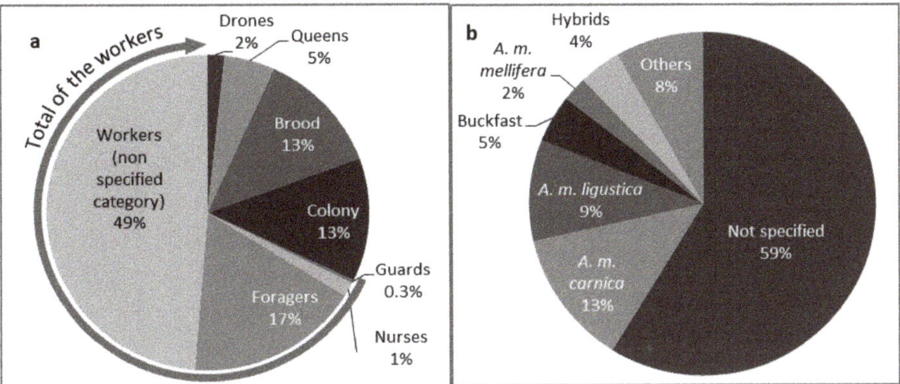

Figure 3. *Apis mellifera* categories (**a**) and subspecies (**b**) studied in the 293 articles related to the impact of stressors on *Apis mellifera* published between 2007 and 2017. "Others" refers to *A. m. Anatoliaca, A. m. caucasica, A. m. iberiensis, A. m. intermissa, A. m. jemenatica, A. m. macedonica, A. m. adansonii, A. m. andersonii, A. m. bandasii, A. m. carpathica, A. m. meda*, Kona bees (Hawaii) and Russian bees.

3.3. *What Stressors Were Studied and Did They Have an Impact on Bee Health?*

Biotic stressors were very seldom studied (11%) compared to abiotic ones (89%).

3.3.1. Biotic Stressors

Publications on biotic stressors (mostly parasitic and infectious agents (PIAs)) mainly concerned the parasitic mite *Varroa destructor* (Mesostigmata: Varroidae), the fungal agent *Nosema spp.* (33% and 32%, respectively) and viruses (17%). These three categories represent the most widespread parasitic and infectious agents in bee colonies (Figure 4a). Predators and the small hive beetle *Aethina tumida* (Coleoptera: Nitidilidae) were very little studied. However, the recent detection of the latter in Europe and the Philippines [31] and the expansion of the Asian hornet *Vespa velutina* (Hymenoptera: Vespidae) could reverse this trend.

3.3.2. Abiotic Stressors

The most studied abiotic stressors were pesticides (61%, Figure 4b). Insecticides were the most tested (half of them were neonicotinoids (Figure 4d)) while fungicides and herbicides were under-studied. Beekeeping practices were relatively highly studied (29%). Three quarters of the beekeeping practices under study were PIA control systems, whether they used chemicals or not (Figure 4c). Bee nutrition was relatively well investigated (17%), while queen management, wintering methods and hive transfers were little studied (≤4%). Among the PIA control methods, "hard" chemical treatment methods [32] were more studied (60%) than "soft" methods. Indeed, these products are often acaricides or fungicides, potentially harmful for bees. Essential oils and organic acids, considered as "soft" methods [32], were studied in 23% and 12% of cases, respectively. Furthermore, Jacques et al. [33] have shown during the EPILOBEE surveillance project that poor beekeeping practices and the lack of expertise of some beekeepers represented one of the major causes of colony loss in Europe. It should be noted that the present meta-analysis only takes into account veterinary products (mostly acaricide treatments) and techniques currently used in beekeeping. Experiments studying any other active ingredients (e.g., toxicity of essential oils not used in beekeeping) were discarded from the analysis.

Other stressors like GMOs, metals (aluminum, lead, cadmium, arsenic and iron through oral exposure), climate and habitat fragmentation were poorly studied (≤4%).

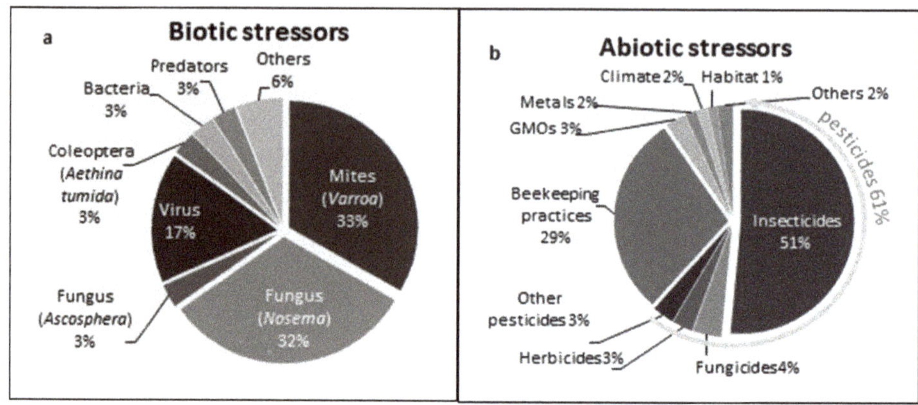

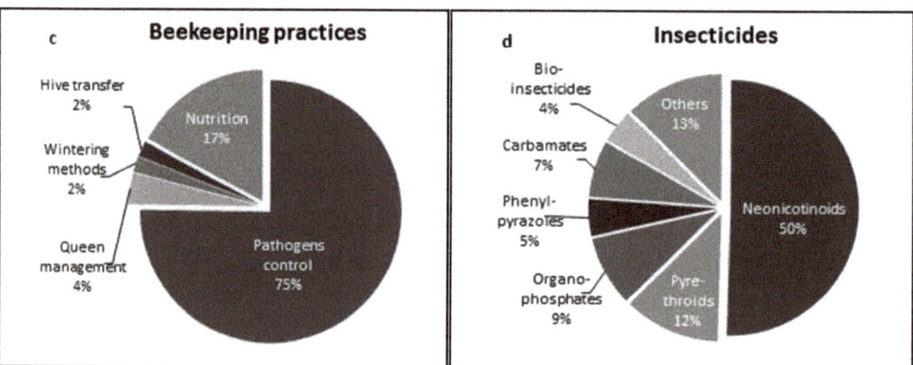

Figure 4. Proportion of the different biotic (**a**) and abiotic (**b**) stressors, beekeeping practices (**c**) and insecticides (**d**), identified in the articles related to the impact of stressors on *Apis mellifera* published between 2007 and 2017. "Other pesticides" (Figure b) refers to adjuvants, inert ingredients, pesticide residues in fields, and wood preservatives. The "other" sections are detailed in Supplementary Materials, Table S2.

3.3.3. Stressors' Impact

In order to determine which stressors affected the most honey bees, Sankey diagrams were generated for each stressor (Figure 5). All the most studied stressors (parasitic and infectious agents, insecticides, chemical veterinary treatments and beekeeping practices other than chemical veterinary treatments) affected the majority of the parameters studied at all scales. There was an exception for veterinary treatments and insecticides, for which about 50% of the parameters studied at the colony level were not impacted. GMOs, although little studied, did not generally have much effect. In particular, results showed no effect on bee mortality and few impacts were observed at the colony or individual level. However, the studied GMOs were mainly bt maize, and as bees are not sensitive to the *bacillus thuringiensis* toxins [34], it is consistent that these GMOs were not demonstrated to have an impact. On the other hand, all the articles dealing with the impact of climate or habitat fragmentation (Figure 5) showed effects, particularly at the colony level. Since the number of publications was very low, the actual effects on bee health should be confirmed by other studies. Exposure to metals appeared to have a significant molecular impact on bees, but very few studies were conducted and none studied colony endpoints. It would therefore be interesting to fill this gap and to relate the molecular effects to the possible effects on the colony.

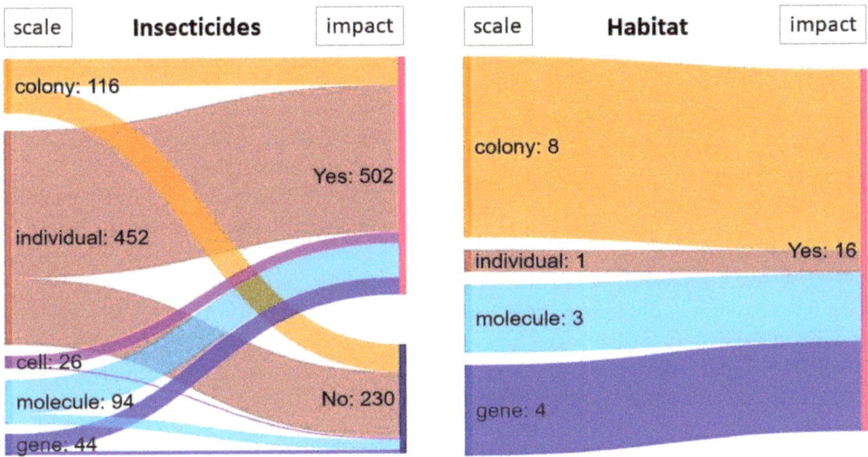

Figure 5. Examples of Sankey diagrams obtained for the stressors "insecticides" and "habitat". The flux represents the number of parameters studied at each biological scale (on the left of the diagram) which have been impacted or not (on the right).

3.3.4. Co-Exposures

Only 20% of the publications investigated interactions between stressors. The most studied were interactions between different pesticides followed by pesticide–PIAs interactions. Other co-exposures included PIAs–nutrition or pesticide–nutrition interactions. The combinations of exposures to be studied is infinite, therefore to help in the decision making process, Henry et al. [35] proposed a procedure to narrow down the panel of options.

3.4. What Methods Were Used in the Articles to Measure Various Endpoints?

Different techniques were used to study the impact of stressors on *A. mellifera* according to the level of biological organization of the measured endpoint: colony, individual, cellular, molecular or genetic endpoints. Individual endpoints were by far the most studied (53%), followed by colony endpoints (21%) and molecular endpoints (14%). Only 8% and 4% of the endpoints were genetic and cellular endpoints, respectively. As colony and individual are the most prevalent endpoints in terms of their relative proportion, we will describe in detail below how these endpoints were produced.

3.4.1. Colony Endpoints

Many authors quantified demographic parameters (number of adult bees, quantity of capped or uncapped brood and/or eggs) or the hive production (number of honey, pollen or nectar combs). In most cases (49%), the number of individuals or combs were estimated by using several similar yet different techniques of frame observation (Figure S5a). Frame or hive weight was also used (13%) to assess brood development and colony size or productivity.

The techniques used to quantify Nosema or Varroa infestation (e.g., samples washed with water or alcohol, powdered sugar, microscopy, sticky boards, etc.) represented 8% of the techniques used at the colony scale. Virus infection was analyzed by PCR (8% of the techniques). The mortality of the entire colony (over the winter or not) was only studied in 7% of the cases. Some authors did not specify which techniques had been implemented (8%). Other methods included moisture and temperature sensors, queen marking to evaluate their renewal, honey extraction techniques, in vitro rearing to study the larvae emergence rate, mathematical models, or field studies that attempt to correlate observed mortality rates with potentially harmful events for bees (i.e., climate events, pesticide use).

3.4.2. Individual Endpoints

Among the parameters studied at the individual level, 56% were non-behavioral parameters.

The most frequent non-behavioral endpoint (Figure S5b) was the mortality rate (73%), which was evaluated with different techniques. Histology techniques (9%) were used to study tissue damage, and in particular, the workers' hypopharyngeal gland ultrastructure. The impact of stressors on the development, immunity or reproduction of honey bees was also assessed.

The behavioral trials (Figure S5c) mainly used the proboscis extension reflex (PER) in conditioning protocols (21%) or not (5%). Observation cages or hives were also used extensively (18% and 6%, respectively), sometimes in association with cameras (video-tracking). The cages were cardboard boxes, Petri dishes or other devices. These devices, as well as other systems often designed by the authors and used to study phototaxy, were used to evaluate abnormal behavior, locomotion, dance and activities in the hive as well as social interactions.

Foraging is an important parameter of behavior, which is mainly studied by counting the foragers in the fields or by recording the number of bees entering and/or exiting the hive. Other methods have been identified but are rarely used, such as pollen traps or weighing of foragers. Marking individuals with color marks or radars (harmonic or radio frequency identification) was used in 11% of cases. Marking was often associated with artificial feeders to study the flight parameters of foragers, or with releasing the honey bees at distance from the hive to test their ability to return.

This inventory revealed a great number of techniques used to study a multitude of parameters. This great diversity of methods may be related to the fact that at present there are only five standard procedures to test chemicals on bees: two acute toxicity tests by ingestion or by contact with adults [36,37], a chronic oral toxicity test with adults [38] and two larval intake toxicity tests [39,40]. A test to evaluate homing success [41] is currently being ring tested for validation, and has not yet been fully accredited. Various tests have been listed [42], but standard tests are still under development and standardization efforts need to be continued.

The diversity of the tests was also related to the large number of parameters that can be evaluated. It is therefore important to identify the most relevant parameters for assessing bee health.

3.5. Did the Evidence of an Impact Vary According to the Type of Study or to the Scale of Study?

In this part of the analysis, we analyzed the impact of the stressor on the endpoint studied (e.g., workers' mortality rate, expression of detoxication genes, immune enzymes' activity, brood capping rate, etc.). The objectives were to determine if the evidence of an impact varied according to the type of study or the scale of study. The impact on the parameter could be "positive" or "negative", but this modality was not recorded.

3.5.1. Type of Study

We investigated whether the parameters studied in the articles were differently affected by a stressor depending on whether the study was carried out in the field (38%) or in a laboratory (48%). In the field, the difference between the number of impacted and non-impacted parameters was very small (Figure 6) yet statistically significant ($p = 0.04$, Chi-square test). In laboratory studies, two-thirds of the parameters were impacted by the stressor and one third was not. This difference could be explained by the effect of stress exposure, by a dose effect—the doses tested in a laboratory may be higher than the doses to which bees are exposed in fields—or by other effects such as co-exposure to multiple stressors and interactions between different products, which are difficult to control in field experiments.

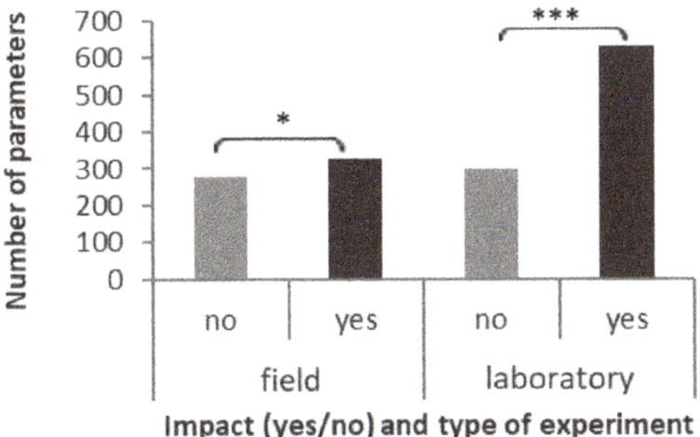

Figure 6. Number of parameters impacted or not by the stressor studied according to the type of experimentation (n = 1532). (* p < 0,05; ** p < 0,01; *** p < 0,001, Chi square test). "Field studies" refer to studies in which the treatment was performed outside, in hives placed in fields or directly in the fields; "laboratory studies" refer to studies in which the treatment was conducted in the laboratory.

3.5.2. Scale of Study

When comparing the results obtained at different scales of study, at the colony scale the difference between the number of impacted and non-impacted parameters was not significant (Figure 7, p = 0.607, Chi-square test). On all other scales, significantly more parameters were demonstrated to be impacted by a stressor than not impacted. This result demonstrated the buffering effect of the colony, which compensates for individual effects.

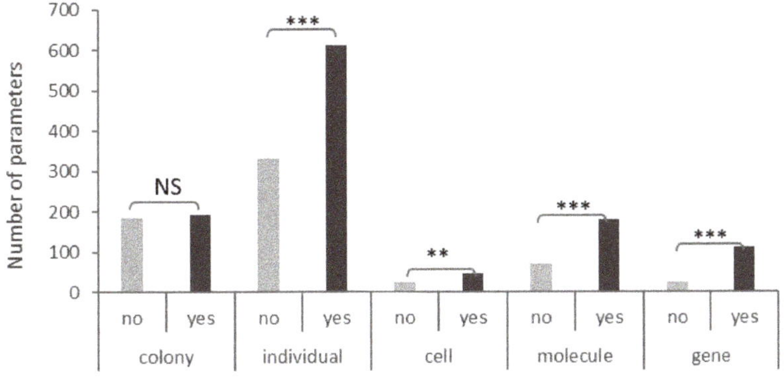

Figure 7. Number of parameters impacted or not by the stressor studied according to the scale of study (n = 1805). (* p < 0,05; ** p < 0,01; *** p < 0,001, NS = Not significant, Chi square test).

Therefore, the same result was observed for parameters studied in the field and parameters studied at the colony level; the difference between impacted and not-impacted parameters was not significant. However, colony endpoint measurements were the main parameters studied in field tests. Therefore, we could not determine whether this effect on the impact of stressors was linked to the type of experimentation or to the buffer effect of the colony. It is very likely that both were involved and other confounding factors may also be responsible for this. Nevertheless, these two effects represent important methodological and biological barriers. Indeed, under natural conditions, it is very difficult to control bees' exposure to stressors and their interactions [9]. In addition, conducting robust studies requires a very large number of replicas, which may pose methodological problems in the field [43]. Mathematical modeling methods might circumvent these technical difficulties.

The buffer effect of the colony is very difficult to take into account in experiments and risk assessment. Indeed, when the individuals of a colony are affected by a stress, if the colony sets up measures to compensate for these individual effects, the impact of the stressor will not be evidenced while the colony is suffering. For example, Henry et al. [12] have shown that when the colony loses its foragers in an abnormally large way, it changes the way the reproductive effort is allocated between the brood of workers and drones: production of males is delayed, while the production of workers is strengthened. The colony size is then maintained, as well as the honey production. Thus, the colony appears to be in good health even though its foragers disappear and the delay in male production may be problematic for mating. This also raises the question of the time scale of an experiment: how long can a colony compensate for a stress without visibly suffering? Are the tests long enough to observe deleterious effects on colonies? It is essential to set up techniques that address these issues. Radio frequency identification (RFID) chips are a first solution since they enable real-time observation of foragers' disappearance.

4. Conclusions

This meta-analysis highlights the great diversity of techniques used by researchers in honey bee experimentations and the need to standardize the protocols. To do so, populations at risk, critical stages of development and the most relevant parameters to be measured should be identified. New standard tests should be developed, especially to better study the sub-lethal effects of stressors on bee health. In addition, greater importance should be given to the bee subspecies studied, to understanding the differences in their sensitivity to stress, and if possible, to identify one or two indicator subspecies. Moreover, this study highlighted the need to break through two important methodological and biological barriers that make risk assessment difficult: the technical difficulties encountered in field tests and the buffer effect of the colony. New technologies such as RFID chips or mathematical modeling could help to overcome these obstacles. This study also highlighted innovative research paths, particularly with regard to the impact of climate and habitat fragmentation, which, according to the few studies already carried out, could have significant deleterious consequences on bee colonies. Finally, as also pointed out by Benuszak et al. [20], efforts to strengthen the number of studies on the impact of co-exposures and metabolites should be continued. In order to develop standard protocols, the search for biomarkers as diagnostic tools seems to be an interesting route of exploration. This biomarker research should be facilitated by "omics" techniques such as genomics, transcriptomics, proteomics, and metabolomics.

Supplementary Materials: The following are available online at http://www.mdpi.com/1424-2818/12/1/7/s1. Figure S1: Search string used to screen the SCOPUS and CAB ABSTRACT databases to select articles related to the impact of stressors on *Apis mellifera* published between 2007 and 2017, Figure S2: list of stressors considered as "anecdotal" to discard related articles from the review on the impact of stressors on *Apis mellifera* published between 2007 and 2017, Table S1: list of the 293 articles related to the impact of stressors on *Apis mellifera* published between 2007 and 2017 included in the analysis, Figure S3: Percentage of articles related to the impact of stressors on *Apis mellifera* ($n = 293$) included in the study and published in scientific journals between 2007 and 2017 (last access to database: March, 6th 2017), Figure S4: Number of publications ($n = 293$) related to the impact of stressors on *Apis mellifera* published between 2007 and 2017 studying the different bee categories according to year of publication, Figure S5: Proportion of the different methods used at the colony scale (a), at the individual scale in non-behavioral trials (b) and in behavioral trials (c), at the cellular (d), the molecular (e) and the genetic (f) scales. ("others": set of methods, each representing 2% or less. See Table S2 for details), Table S2: Detail of the "Others" sections of the Figure 4 and Figure S3.

Author Contributions: M.-P.C. had the original idea for the study. T.H. and M.-P.C. produced the methodology. T.H. realized the meta-analysis of papers, the database production, the data analysis and the initial version of the paper. M. -P.C. and M.L. reviewed the paper. All authors have read and agreed to the published version of the manuscript.

Funding: This research received no external funding to ANSES.

Acknowledgments: The authors wish to thank Druesne and Guitton for their help in bibliography management.

Conflicts of Interest: The authors declare no conflict of interest.

References

1. Biesmeijer, J.C.; Roberts, S.P.M.; Reemer, M.; Ohlemüller, R.; Edwards, M.; Peeters, T.; Schaffers, A.P.; Potts, S.G.; Kleukers, R.; Thomas, C.D.; et al. Parallel Declines in Pollinators and Insect-Pollinated Plants in Britain and the Netherlands. *Science* **2006**, *313*, 351–354. [CrossRef] [PubMed]
2. Klein, A.-M.; Vaissière, B.E.; Cane, J.H.; Steffan-Dewenter, I.; Cunningham, S.A.; Kremen, C.; Tscharntke, T. Importance of pollinators in changing landscapes for world crops. *Pro. Biol. Sci.* **2007**, *274*, 303–313. [CrossRef] [PubMed]
3. Decourtye, A.; Devillers, J. Ecotoxicity of neonicotinoid insecticides to bees. *Adv. Exp. Med. Biol.* **2010**, *683*, 85–95. [PubMed]
4. Allen-Wardell, G.; Bernhardt, P.; Bitner, R.; Burquez, A.; Buchmann, S.; Cane, J.; Cox, P.; Dalton, V.; Feinsinger, P.; Ingram, M.; et al. The Potential Consequences of Pollinator Declines on the Conservation of Biodiversity and Stability of Food Crop Yields. *Conserv. Biol.* **1998**, *12*, 8–17.
5. Vanengelsdorp, D.; Meixner, M.D. A historical review of managed honey bee populations in Europe and the United States and the factors that may affect them. *J. Invertebr. Pathol.* **2010**, *103* (Suppl. 1), S80–S95. [CrossRef] [PubMed]
6. Benaets, K.; Van Geystelen, A.; Cardoen, D.; De Smet, L.; de Graaf, D.C.; Schoofs, L.; Larmuseau, M.H.D.; Brettell, L.E.; Martin, S.J.; Wenseleers, T. Covert deformed wing virus infections have long-term deleterious effects on honeybee foraging and survival. *Proc. Biol. Sci.* **2017**, *284*, 20162149. [CrossRef]
7. Alburaki, M.; Cheaib, B.; Quesnel, L.; Mercier, P.; Chagnon, M.; Derome, N. Performance of honeybee colonies located in neonicotinoid-treated and untreated cornfields in Quebec. *J. Appl. Entomol.* **2016**, *141*, 112–121. [CrossRef]
8. Fine, J.D.; Cox-Foster, D.L.; Mullin, C.A. An Inert Pesticide Adjuvant Synergizes Viral Pathogenicity and Mortality in Honey Bee Larvae. *Sci. Rep.* **2017**, *7*, 1–9. [CrossRef]
9. Goulson, D.; Nicholls, E.; Botías, C.; Rotheray, E.L. Bee declines driven by combined stress from parasites, pesticides, and lack of flowers. *Science* **2015**, *347*, 125–957. [CrossRef]
10. Christen, V.; Bachofer, S.; Fent, K. Binary mixtures of neonicotinoids show different transcriptional changes than single neonicotinoids in honeybees (*Apis mellifera*). *Environ. Pollut.* **2017**, *220*, 1264–1270. [CrossRef]
11. Le Conte, Y.; Navajas, M. Climate change: Impact on honey bee populations and diseases. *Rev.-Off. Int. Epizoot.* **2008**, *27*, 485–497. [PubMed]
12. Henry, M.; Cerrutti, N.; Aupinel, P.; Decourtye, A.; Gayrard, M.; Odoux, J.-F.; Pissard, A.; Rüger, C.; Bretagnolle, V. Reconciling laboratory and field assessments of neonicotinoid toxicity to honeybees. *Proc. Biol. Sci.* **2015**, *282*, 20152110. [CrossRef] [PubMed]
13. Schneider, S.; Eisenhardt, D.; Rademacher, E. Sublethal effects of oxalic acid on *Apis mellifera* (Hymenoptera: Apidae): Changes in behaviour and longevity. *Apidologie* **2012**, *43*, 218–225. [CrossRef]
14. Aliouane, Y.; El Hassani, A.K.; Gary, V.; Armengaud, C.; Lambin, M.; Gauthier, M. Subchronic exposure of honeybees to sublethal doses of pesticides: Effects on behavior. *Environ. Toxicol. Chem.* **2009**, *28*, 113–122. [CrossRef]
15. Bonnafé, E.; Alayrangues, J.; Hotier, L.; Massou, I.; Renom, A.; Souesme, G.; Marty, P.; Allaoua, M.; Treilhou, M.; Armengaud, C. Monoterpenoid-based preparations in beehives affect learning, memory, and gene expression in the bee brain. *Environ. Toxicol. Chem.* **2017**, *36*, 337–345. [CrossRef]
16. Martín-Hernández, R.; Higes, M.; Sagastume, S.; Juarranz, Á.; Dias-Almeida, J.; Budge, G.E.; Meana, A.; Boonham, N. Microsporidia infection impacts the host cell's cycle and reduces host cell apoptosis. *PLoS ONE* **2017**, *12*, e0170183. [CrossRef]
17. Christen, V.; Mittner, F.; Fent, K. Molecular Effects of Neonicotinoids in Honey Bees (*Apis mellifera*). *Environ. Sci. Technol.* **2016**, *50*, 4071–4081. [CrossRef]
18. Pisa, L.W.; Amaral-Rogers, V.; Belzunces, L.P.; Bonmatin, J.M.; Downs, C.A.; Goulson, D.; Kreutzweiser, D.P.; Krupke, C.; Liess, M.; McField, M.; et al. Effects of neonicotinoids and fipronil on non-target invertebrates. *Environ. Sci. Pollut. Res.* **2015**, *22*, 68–102. [CrossRef]
19. EFSA (European Food Safety Authority). Towards an integrated environmental risk assessment of multiple stressors on bees: Review of research projects in Europe, knowledge gaps and recommendations. *EFSA J.* **2014**, *12*, 3594.

20. Benuszak, J.; Laurent, M.; Chauzat, M.-P. The exposure of honey bees (*Apis mellifera*; Hymenoptera: Apidae) to pesticides: Room for improvement in research. *Sci. Total Environ.* **2017**, *587*, 423–438. [CrossRef]

21. Johnson, R.M.; Dahlgren, L.; Siegfried, B.D.; Ellis, M.D. Effect of in-hive miticides on drone honey bee survival and sperm viability. *J. Apic. Res.* **2013**, *52*, 88–95. [CrossRef]

22. Alaux, C.; Folschweiller, M.; McDonnell, C.; Beslay, D.; Cousin, M.; Dussaubat, C.; Brunet, J.-L.; Le Conte, Y. Pathological effects of the microsporidium Nosema ceranae on honey bee queen physiology (*Apis mellifera*). *J. Invertebr. Pathol.* **2011**, *106*, 380–385. [CrossRef] [PubMed]

23. Rangel, J.; Tarpy, D.R. The combined effects of miticides on the mating health of honey bee (*Apis mellifera* L.) queens. *J. Apic. Res.* **2015**, *54*, 275–283. [CrossRef]

24. Williams, G.R.; Troxler, A.; Retschnig, G.; Roth, K.; Yañez, O.; Shutler, D.; Neumann, P.; Gauthier, L. Neonicotinoid pesticides severely affect honey bee queens. *Sci. Rep.* **2015**, *5*, 1–8. [CrossRef]

25. Chaimanee, V.; Evans, J.D.; Chen, Y.; Jackson, C.; Pettis, J.S. Sperm viability and gene expression in honey bee queens (*Apis mellifera*) following exposure to the neonicotinoid insecticide imidacloprid and the organophosphate acaricide coumaphos. *J. Insect Physiol.* **2016**, *89*, 1–8. [CrossRef]

26. Dussaubat, C.; Maisonnasse, A.; Crauser, D.; Tchamitchian, S.; Bonnet, M.; Cousin, M.; Kretzschmar, A.; Brunet, J.-L.; Le Conte, Y. Combined neonicotinoid pesticide and parasite stress alter honeybee queens' physiology and survival. *Sci. Rep.* **2016**, *6*, 31430. [CrossRef]

27. Kairo, G.; Provost, B.; Tchamitchian, S.; Ben Abdelkader, F.; Bonnet, M.; Cousin, M.; Sénéchal, J.; Benet, P.; Kretzschmar, A.; Belzunces, L.P.; et al. Drone exposure to the systemic insecticide Fipronil indirectly impairs queen reproductive potential. *Sci. Rep.* **2016**, *6*, 31904. [CrossRef]

28. Sturup, M.; Baer-Imhoof, B.; Nash, D.R.; Boomsma, J.J.; Baer, B. When every sperm counts: Factors affecting male fertility in the honeybee *Apis mellifera*. *Behav. Ecol.* **2013**, *24*, 1192–1198. [CrossRef]

29. Suchail, S.; Guez, D.; Belzunces, L. Characteristics of imidacloprid toxicity in two *Apis mellifera* subspecies. *Environ. Toxicol. Chem.* **2000**, *19*, 1901–1905. [CrossRef]

30. Rinkevich, F.D.; Margotta, J.W.; Pittman, J.M.; Danka, R.G.; Tarver, M.R.; Ottea, J.A.; Healy, K.B. Genetics, Synergists, and Age Affect Insecticide Sensitivity of the Honey Bee, *Apis mellifera*. *PLoS ONE* **2015**, *10*, e0139841. [CrossRef]

31. Giangaspero, M.; Turno, P. Aethina tumida, an Exotic Parasite of Bees. *Clin. Microbiol.* **2015**, *4*, 1000e128. [CrossRef]

32. Rosenkranz, P.; Aumeier, P.; Ziegelmann, B. Biology and control of *Varroa destructor*. *J. Invertebr. Pathol.* **2010**, *103* (Suppl. 1), S96–S119. [CrossRef]

33. Jacques, A.; Laurent, M.; Ribière-Chabert, M.; Saussac, M.; Bougeard, S.; Budge, G.E.; Hendrikx, P.; Chauzat, M.-P. A pan-European epidemiological study reveals honey bee colony survival depends on beekeeper education and disease control. *PLoS ONE* **2017**, *12*, e0172591. [CrossRef]

34. Yi, D.; Fang, Z.; Yang, L. Effects of Bt cabbage pollen on the honeybee *Apis mellifera* L. *Sci. Rep.* **2018**, *8*, 1–6. [CrossRef]

35. Henry, M.; Becher, M.A.; Osborne, J.L.; Kennedy, P.J.; Aupinel, P.; Bretagnolle, V.; Brun, F.; Grimm, V.; Horn, J.; Requier, F. Predictive systems models can help elucidate bee declines driven by multiple combined stressors. *Apidologie* **2017**, *48*, 328–339. [CrossRef]

36. *OECD Test No. 213: Honeybees, Acute Oral Toxicity Test*; OECD: Paris, France, 1998.

37. *OECD Test No. 214: Honeybees, Acute Contact Toxicity Test*; OECD: Paris, France, 1998.

38. *OECD Test No. 245: Honey Bee (Apis mellifera L.), Chronic Oral Toxicity Test (10-Day Feeding)*; OECD: Paris, France, 2017.

39. *OECD Test No. 237: Honey Bee (Apis mellifera) Larval Toxicity Test, Single Exposure*; OECD: Paris, France, 2013.

40. *OECD Series on Testing & Assessment No. 239: Guidance Document on Honey Bee (Apis mellifera) Larval Toxicity Test following Repeated Exposure*; OECD: Paris, France, 2016.

41. Henry, M.; Béguin, M.; Requier, F.; Rollin, O.; Odoux, J.-F.; Aupinel, P.; Aptel, J.; Tchamitchian, S.; Decourtye, A. A common pesticide decreases foraging success and survival in honey bees. *Science* **2012**, *336*, 348–350. [CrossRef]

42. Dietemann van der Zee, V.; Ellis, J.; Neumann, J. The COLOSS BEEBOOK, Volume I: Standard methods for *Apis mellifera* research. *J. Apic. Res.* **2013**, *52*. [CrossRef]
43. Woodcock, B.A.; Heard, M.S.; Jitlal, M.S.; Rundlöf, M.; Bullock, J.M.; Shore, R.F.; Pywell, R.F. Replication, effect sizes and identifying the biological impacts of pesticides on bees under field conditions. *J. Appl. Ecol.* **2016**, *53*, 1358–1362. [CrossRef]

 *diversity*

Review

Vespa velutina: An Alien Driver of Honey Bee Colony Losses

Daniela Laurino *, Simone Lioy, Luca Carisio, Aulo Manino and Marco Porporato

Department of Agricultural, Forest and Food Sciences, University of Turin, 10095 Grugliasco (Turin), Italy; simone.lioy@unito.it (S.L.); luca.carisio@unito.it (L.C.); aulo.manino@unito.it (A.M.); marco.porporato@unito.it (M.P.)

* Correspondence: daniela.laurino@unito.it

Received: 15 November 2019; Accepted: 18 December 2019; Published: 20 December 2019

Abstract: *Vespa velutina*, or Asian yellow-legged hornet, was accidentally introduced from China to other parts of the world: South Korea in 2003, Europe in 2004, and Japan in 2012. *V. velutina* represents a serious threat to native pollinators. It is known to be a fierce predator of honey bees, but can also hunt wild bees, native wasps, and other flying insects. When *V. velutina* colonies are developed, many hornets capture foraging bees which are coming back to their hives, causing an increase in homing failure and paralysis of foraging thus leading to colony collapse. The hornets may enter weak beehives to prey on brood and pillage honey. Unlike *Apis cerana*, *Apis mellifera* is unable to cope with the predation pressure of *V. velutina*. Monitoring the spread of an invasive alien species is crucial to plan appropriate management actions and activities to limit the expansion of the species. In addition, an early detection of *V. velutina* in areas far away from the expansion front allows a rapid response aimed to remove these isolated populations before the settlement of the species. Where *V. velutina* is now established, control measures to prevent colony losses must be implemented with an integrated pest management approach.

Keywords: *Vespa velutina*; alien driver; honey bee; damage; pollinator

1. Introduction

Invasive alien species have always been a risk to ecosystems. They are a serious obstacle to the conservation of biodiversity, both globally and locally, as their stabilization and spread in new environments break the pre-existing balances. By coming into contact with a new environment, alien species can lead to a gradual degradation and alteration of the new habitat and the decline of indigenous species, until in some cases some of them become extinct [1–3].

The Asian yellow-legged hornet (*Vespa velutina nigrithorax* Du Buysson) is a social wasp, belonging to one of the 11 subspecies [4–6] of *V. velutina* originally present in Continental Asia [7], where it is native to subtropical and temperate areas of Indo-China [6,8].

The species established itself in non-native countries such as South Korea in 2003 [9] in the southern port town of Busan and Japan, on Tsushima Island in 2012 [10,11], in Kitakyushu City on Kyushu Island in 2015 [12] and on Iki Island in 2017 [13]. Arrived in France probably in 2004 along with garden pots imported from China [14,15], the species spread to neighboring countries. From France it reached the Navarra province and Basque country (Spain) in 2010 [16,17], Galicia [18] and Catalunya [19] in 2012, Majorca Island (Spain) in 2015 [20,21], the Minho province (Portugal) in 2011 [22], and Flobecq in the Hainaut province (Belgium) in 2011 [23]. In 2012, the Asian yellow-legged hornet was detected for the first time in Italy in the Liguria Region [24]; afterwards the hornet started to spread in this region mainly along the coastline [25–27]. In Piedmont Region (Italy) arrived in 2013 [25,26]. It was detected also in Veneto and Lombardy Regions (Italy) between 2016 and 2017, with no more reports in the following years, and in Tuscany Region (Italy) in 2017. By 2017, the species had colonized an area of at least 1,110 km^2 in Italy [27]. *V. v. nigrithorax* was firstly recorded in Germany in 2014 and a nest was found in Büchelberg (Rheinland-Pfalz) [28]. In 2016, few hornets were found in the United Kingdom, and in 2017 also in Netherland, Switzerland [29], and Scotland [30], Figure 1.

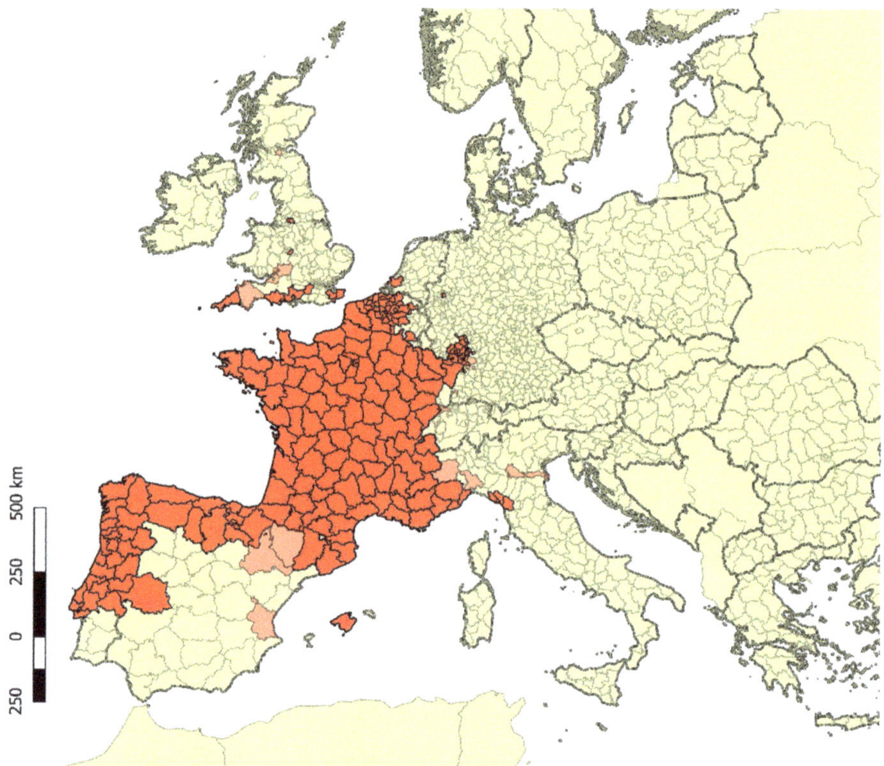

Figure 1. Presence of *V. v nigrithorax* in Europe estimated from several sources. Red areas indicate districts where hornets are established or have been reported in 2018 and 2019. Light-red areas show districts where hornets or nests have been exclusively spotted in the past until year 2017.

The spread of *V. v. nigrithorax* in Europe and in non-native Asian regions seems to respect the predicted climatic suitability maps modelled by Villemant et al. [31]. Global warming could worsen the current situation [32]. This hypothesis was confirmed by Rodríguez-Flores et al. [18]: High minimum temperatures, dew temperature, relative humidity and low maximum temperatures favor the occurrence and spread of *V. v. nigrithorax*. These conditions are common in coastal areas and can promote the rapid dispersal of this pest.

V. v. nigrithorax creates considerable damage to the environment and beekeeping activities. For this reason, the species has been included by European Union in the black-list of invasive alien species (Reg. EU 1141/2016) for which it is mandatory to develop surveillance plans and actions to limit its spread as well as control and containment strategies. The Japanese Ministry of the Environment added this hornet to the list of invasive alien species in 2015 [33].

2. Biology

The colony of *V. v. nigrithorax* is started by a single inseminated queen that builds, using fibrous substances of plants origin and saliva, a primary nest after overwintering, typically in April, thus producing the first workers. During the warm season, they enlarge the primary nest (which has an approximate size between 4 cm and 15 cm) directly or build a secondary nest normally on treetops [18]. Nests have normally a circular shape and can grow up to 100 cm in diameter, containing several thousands of hornets. Rome et al. [34] report up to 13,300 adults and 563 new queens from a single nest. At the end of the summer, reproductive individuals emerge and mate; the colonies generally collapse in late autumn or winter, while newly-mated queens search for a place where they can overwinter and, the following year, they start a new cycle [8,27,34,35], Figure 2.

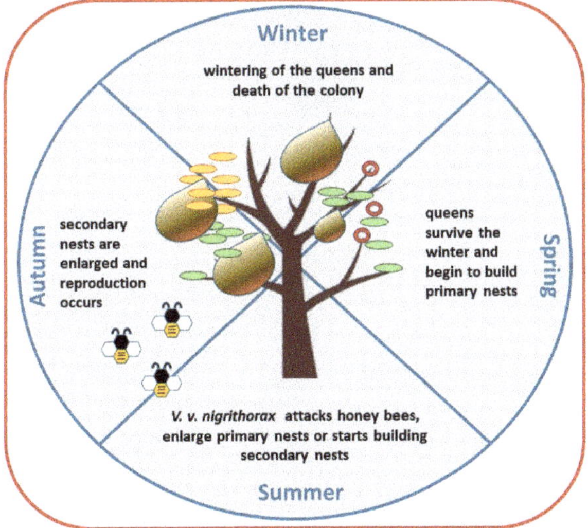

Figure 2. Life cycle of *V. v. nigrithorax*.

Hornets use olfactory stimuli to search for long-distance food sources, especially with regard to the localization of honey bee colonies, but the nature of these stimuli is not yet entirely clear [36–38].

The components of the hive, which attract the most attention of *V. v. nigrithorax*, have been the subject of study for some years. Hornets are strongly attracted by the odour of some hive products, especially pollen and honey [37]. A laboratory study showed that *V. v. nigrithorax* workers can use both visual and olfactory cues to locate honey bees [39].

Honey bees attract the attention of the hornets thanks to the production of geraniol, component of the aggregation pheromone of the colony. Less effective than pollen and honey, but still attractive is the royal jelly, thanks to the presence of homovanillyl alcohol (HVA) and methyl-4hydrobenzoate (HOB), substances that are part of the pheromone produced by the honey bee queens, but also present in the royal jelly. Betaocimene emitted by larvae also produces olfactory stimuli that are very attractive to hornets [37].

The pheromones produced by the hornet colony's components are being studied for their possible use in biological control techniques [40]. Couto et al. [41], in a neurobiological works on *V. v. nigrithorax*, showed the presence of several microstructures in the antennal lobe of the males, which are probably linked to sex pheromones. Recently Wen et al. [42] announced the isolation of the sex pheromones from *V. velutina* queens. Cheng et al. [43] proved that *V. velutina* uses sting venom volatiles as an alarm pheromone.

3. Impacts

In Europe, *V. v. nigrithorax* is considered invasive, both for its expansion capabilities at European scale [44–46] and the impacts that it could produce by preying on honey bees and native insects [35,47–49]. The species can cause serious damage and imbalances to biodiversity and ecosystems in areas where it has been introduced. This is aided by *V. v. nigrithorax*'s high reproductive rate, high dispersal ability, broad diet, wide habitat preference, superior competitive ability, and most importantly, multiple mating of its queens [13,50,51].

Although the species is not considered in Europe more dangerous than native hornets and wasps [52], it can cause problems to human-health and several accidents and some fatal events were recorded [53]. Moreover, by frequently establishing colonial nests in urban areas, *V. v. nigrithorax* could generate social impacts due to citizens' perception of fear of possible stings [54–56]. *V. v. nigrithorax* causes multiple threats, however the extent of the impacts produced has not yet been documented and quantified exhaustively. Despite this lack, it is possible to assess the components and/or the activities that are affected or may be impacted by this hornet. From the analysis of the research conducted so far in Europe it is possible to reckon three major negative impacts that the species may create:

1. Economic threats: Loss of honey bee colonies and decreasing of beekeeping products; cost for control activities and nests removal;
2. Ecological threats: Impacts on biodiversity; impacts on pollination activity; competition with native species;
3. Public health threats: Potential risk for citizens.

3.1. Damage to Beekeeping

The hornets hunt foraging honey bees returning to their colonies by hovering in front of the hive entrance, grabbing the honey bees in flight, and killing them with their jaws. The thorax is then selected, for the high protein content provided by the flight muscles, and transported to the nest to be fed to the larvae.

The intense predatory activity of *V. v. nigrithorax* towards honey bees can generate a decrease in the strength of the honey bees colonies and the subsequent collapse of the family. This is due to the increase in the number of foraging bees which do not come back in their hive with a consequent increase of the overall probability of homing failure and finally with the disruption of colony foraging activities, which leads to complete foraging paralysis [57]. This primarily generates economic damage to the beekeeping sector, as well as a decrease in the number of honey bees in the environment, resulting in a decline in the ecosystem pollination service.

In some European regions, predation of *V. v. nigrithorax* has resulted in the loss of almost 50% of bee hives. In the south-west of France, beekeepers reported losses of between 30% and 80% of honey bee families, resulting in poor production of honey and other beehive products. In 2010 in Gironde (France), due to *V. v. nigrithorax*, the Union Nationale pour l'Apiculture Francaise declared that 30% of bee hives were destroyed or weakened [58]. In Western Liguria (Italy) the authors of the present review (unpublished results) have found an increase of 18% in winter colony losses in areas where *V. v. nigrithorax* is not controlled.

The costs incurred both for the implementation of public information campaigns and for the destruction of *V. v. nigrithorax* nests are relevant economic issues. In 2011 in France, the beekeepers' organization Groupement de Dèfense Sanitaire des Abeilles (GDSA) coordinated the destruction of more than 1,000 nests in Aquitania, while a private company destroyed about 500 nests in the Toulouse area. The total cost of these interventions can be quantified to more than 165,000 euros.

Leza et al. [21] demonstrated that the presence of *V. v. nigrithorax* produces an increase of oxidative stress in honey bee workers under field conditions. This leads to a higher expression and activity of antioxidant enzymes and mitochondrial-related genes and higher lipid oxidative damage in the individuals of the colony exposed to this predator. Other authors reported that other stressors, like herbicides or migratory management, could increase lipid peroxidation in honey bees [59] suggesting that these situations, along with the presence of *V. v. nigrithorax*, could affect honey bees' health [21].

The apiaries are a very attractive source of food for *V. v. nigrithorax*, because there is a high concentration of honey bees. Studies performed in France have demonstrated that in urbanized environments, where the concentration of apiaries is high, the diet of *V. v. nigrithorax* is composed for almost 70% of honey bees and other similar species (Apoidea) [60]. Monceau et al. [49] monitored the predation of *V. v. nigrithorax* on apiaries. In an apiary with six beehives, in the sampling period they caught a total of 360 workers, and most of these visited the apiary daily. This indicates that once the species has identified an important protein source such as an apiary, it visits the site every day, probably because of a greater success of predation. Of the six beehives monitored in the season, one was completely destroyed, while in the other five the size of the colonies halved. In addition, five *V. v. nigrithorax* nests were discovered within 1 km from the experimental apiary; so it is likely that apiaries were attacked by individuals from different colonies.

The colonies of *Apis mellifera* manifest a certain defensive ability towards *Vespa crabro*, the European native hornet, but fail to implement effective defensive behaviours towards *V. v. nigrithorax*, having had no opportunity to co-evolve with this predator. In the Asian regions, where the local bee species, *Apis cerana*, has co-evolved with *V. v. nigrithorax*, the honey bees have instead developed very effective defense techniques, resulting in suffocation and heating of the predator (balling), and the formation of a compact agglomeration of honey bees on the flying board [61]. *A. mellifera* is also able to curl up hornets that rest on the flying board of hives, but this behavior does not reach large percentages of success.

At the end of summer, when the colonies of *V. v. nigrithorax* are very populous, hornets can get to besiege the hives and penetrate inside them annihilating the colonies of honey bees. The use of doors with passages less than 5.5 mm in diameter can prevent the entrance of hornets and delay the definitive collapse of the colonies, but if the beekeeper does not intervene to eliminate hornets, the honey bees cannot get out and the colony is destined to collapse (Figure 3).

Figure 3. *V. v. nigrithorax* in hunting activity in front of a hive (left) and heavy attack of hornets on the flying board of a hive (right).

As it often happens in many species of insects, climatic conditions, especially temperature and humidity, affect the predator's activities. In the case of *V. v. nigrithorax*, the increase in the efficiency of predation, which is most evident in the middle hours of the day, would be the result of an increase in temperatures and the level of solar radiation [38].

3.2. Impact on Ecosystem

The predatory activity of *V. v. nigrithorax* has a negative impact on insect communities, reducing their abundance and may cause damage to local biodiversity even at the ecosystem level. The predation pressure known in apiaries since the month of July may have similar effects also on other pollinating insects, creating a decline in pollination effectiveness. In fact, besides honey bees, *V. v. nigrithorax* preys on other Hymenoptera, including different species of wild bees and other Vespidae (wasps in general), but also Diptera (flies and mosquitoes), butterflies species and other insects. Species preyed upon by *V. v. nigrithorax* and their proportion varies according to the prey availability in the environment. A French study showed that in an urban environment, *V. v. nigrithorax* preys mostly honey bees and other Apoidea (66% of the diet), while in a woodland environment, bees and other Apoidea drop to 33% and Diptera increase to 32% [60] (Figure 4).

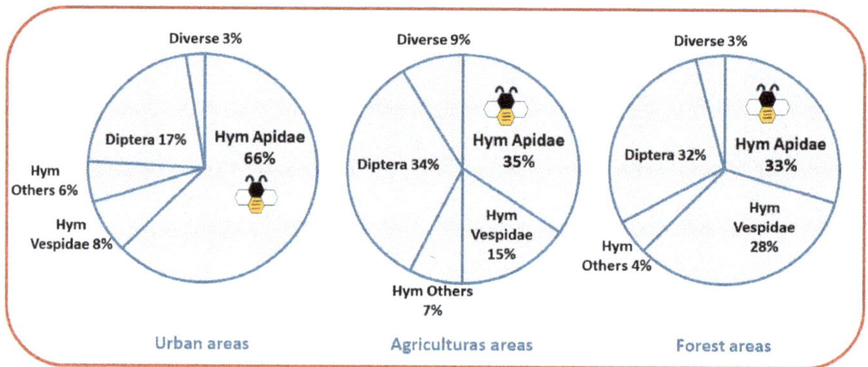

Figure 4. *V. v. nigrithorax* prey spectrum: Preliminary results in three different environments [60].

4. Monitoring and Surveillance Systems

Monitoring the spread of an invasive alien species is crucial to plan appropriate management actions and activities to limit its expansion. Only monitoring and surveillance strategies permit to assess the presence of the species on the territory and identify the areas of expansion or new invasive outbreaks. Since *V. v. nigrithorax* is particularly attracted by honey bees, it is important to involve beekeepers and beekeeper associations to maximize the efficacy of monitoring strategies, together with the contribution of all interested citizens.

An early detection of *V. v. nigrithorax* in areas far away from the expansion front allows to perform a rapid response aimed to remove these isolated populations before the settlement of the species [62]. In fact, *V. v. nigrithorax* queens might be accidentally transported by human activities in very remote areas, where these insects can give rise to new colonies and populations [26]. An early warning and rapid response system (EWRRS) for *V. v. nigrithorax* is based on three key moments (Figure 5).

Results of EWRRS are the rapid detection of the species and the readiness of intervention, which increases the probability of destroying the colonies before the birth and the mating of the future founder queens. This increases the probability of success in the containment of *V. v. nigrithorax*. Key aspects of an effective management strategy are: Simplicity of the procedures; rapid intervention; exportability on a national and international scale; economic sustainability.

Different monitoring methods for *V. v. nigrithorax* exist: Direct observations of hornets in apiaries or on flowers and the use of traps.

Several trap models have been proposed to catch *V. v. nigrithorax* adults; basically, they can be reduced to bottle, funnel, and sticky traps or to a combination of them.

In addition to proprietary baits, many types of self-produced carbohydrate or protein baits can be used. Sugar based baits include beer, vinegar, grenadine, acetic acid, fermented honeycomb juice, honey, different type of sweeteners mixtures, etc. [40]. They are better used between February and May, so to catch the founder queens when they begin the construction of the primary nests, and from August until November, to detect the presence of the species in new areas or to catch the reproductive adults. Demichelis et al. [24] recommend the use of lager beer (0.33 litres with 4.7% alcohol), because it is attractive for the hornets, inexpensive, and selective towards honey bees (Figure 6).

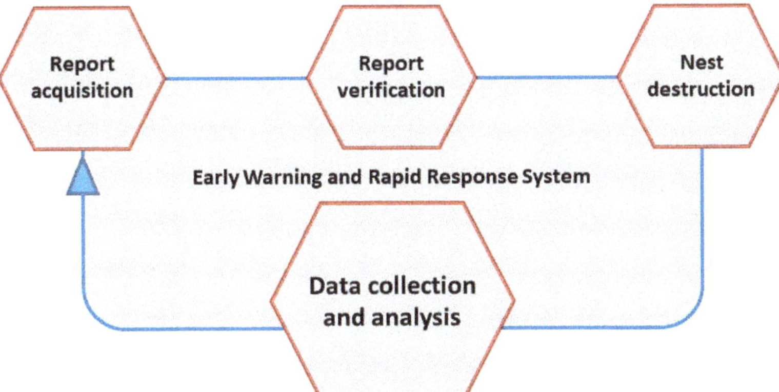

Figure 5. Early warning and rapid response system for *V. v. nigrithorax*.

Figure 6. Bottle trap with lager beer [24].

During colony development, between June and August, protein baits (meat, fish, etc.) can be used. In addition, Rodríguez-Flores et al. [18] highlighted that elevation and meteorological factors influence the effectiveness of bait trapping.

5. Control

Control invasive alien species is difficult and expensive in the long term; therefore, every effort should be undertaken to prevent their establishment and diffusion in new areas. The many methods developed to control *V. v. nigrithorax* in Europe and in Asia outside of its native range have been previously reviewed [9,33,35,40,47].

Until now, no single control method has proved to be fully effective, but the coordinated use of several methods under an integrated pest management approach should greatly reduce the impact of *V. v. nigrithorax* on honey bees and on the environment (Figure 7).

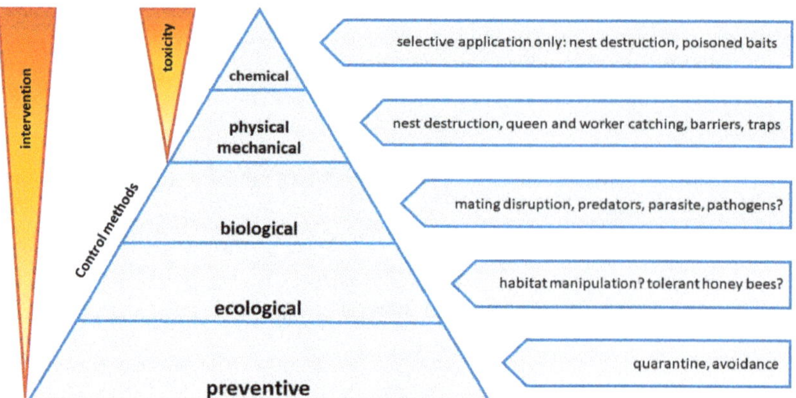

Figure 7. Integrated pest management pyramid showing available or possible methods to control *V. v. nigrithorax*, from the low intervention and not toxic preventive methods to the high impact chemical methods that implies the use of toxic insecticides.

Control efforts may implemented against adult hornets or nests. In the first case, the baited traps used for monitoring purposes can be used either for controlling purposes, but traps are at the moment not selective enough to prevent extensive captures of non-target insects, with possible extensive impacts to native species [57,63]. Other types of control techniques are used or tested in Europe: Bucket poisoned baits, passive traps, electric traps, electric harps, badminton rackets, beehive muzzles, nest gunshot, or the use of the hornet workers as poison carriers [40].

The detection and destruction of the nests of *V. v. nigrithorax* is currently the most effective control method, especially when the nests themselves are destroyed before the reproductive phase of the colony which normally occurs in early September. In any case, it is important to search for and destroy active nests in all stages from foundation to winter [53]. Once a nest has been located, it must be destroyed in a complete way, paying special attention to killing the queen, the majority of the workers, and all the brood present in the combs.

The discovered nests should be immediately destroyed by people with specific training and equipped with suitable personal protective clothes and the necessary tools. The methods of intervention are various in relation to the place where the nests are built and the size of the colonies.

Nests are generally treated with insecticides for hornets and wasps, using also special extendable rods capable of reaching nests that are in high positions (Figure 8).

The control of the populations of *V. v. nigrithorax* is hindered by the difficulty of finding all nests. Embryo nests are small and difficult to observe; later the nests, although very voluminous, are often difficult to be located since covered by tree canopy. In fact, *V. v. nigrithorax* can build nests in several environments, such as natural, rural, and urban areas, and on different substrates [18]. They can be found on trees, shrubs, roofs or balconies of houses, inside gaps, but also in soil cavities or on rocky substrates (Figure 9).

Figure 8. Different methods to destroy the nests depending on their position.

Figure 9. Primary nest built under a roof canopy (left) and secondary nest (right) built on *Acacia dealbata* tree in Liguria Region (Italy).

5.1. Nest Detection

Several techniques are currently available to locate hornet nests, but they are all tedious, extremely labor consuming, and/or expensive [40].

The triangulation involves capturing at least three specimens and their subsequent release from various locations, in order to recording the direction of their flight; if the hornets tend to return in a straight direction, there is a good chance that the three directions they took would intersect at a point that will correspond to nest position [20,64]. The on-view tracking of tagged hornets, which are made more evident with a feather or a cotton, has been suggested [65]. The drone-assisted nest tracking, a theoretical study that uses several drones equipped with cameras capable of analysis an image of a hornet marked with a thread carrying a fluorescent Styrofoam ball, has recently been published [66].

Alternatively to these observational methods, it is possible locate the nests by using equipment developed in recent years.

The infrared thermal imaging camera has been tested in Portugal, in UK [67,68], and in Italy [69].

Radio-telemetry has been implemented in UK as a tool for tracking hornets back to their nests and providing an efficient mean of finding nests in complex environments [68].

A European LIFE project (LIFE STOPVESPA) has just been completed to contain the spread of *V. v. nigrithorax* in Italy by implementing an EWRRS (https://www.vespavelutina.eu/en-us/). For this reason, two prototypes of harmonic entomological radars able to track the flight of hornets in real time and quickly locate the position of nests were developed by the Polytechnic University of Turin (Italy). The radars are capable to follow the flight of hornets equipped with a passive transponder (tag). The radar emits a series of short pulses at a given frequency (9.41 GHz), which are then retransmitted at a double frequency (for this reason, it is called 'harmonic') by the tag fixed on the thorax of the hornets (Figure 10) [70,71]. The tracks obtained allow to follow the path of the hornets and locate the nests. Thanks to the high transmission power, a wide maximum operating distance of 490 m was obtained [72].

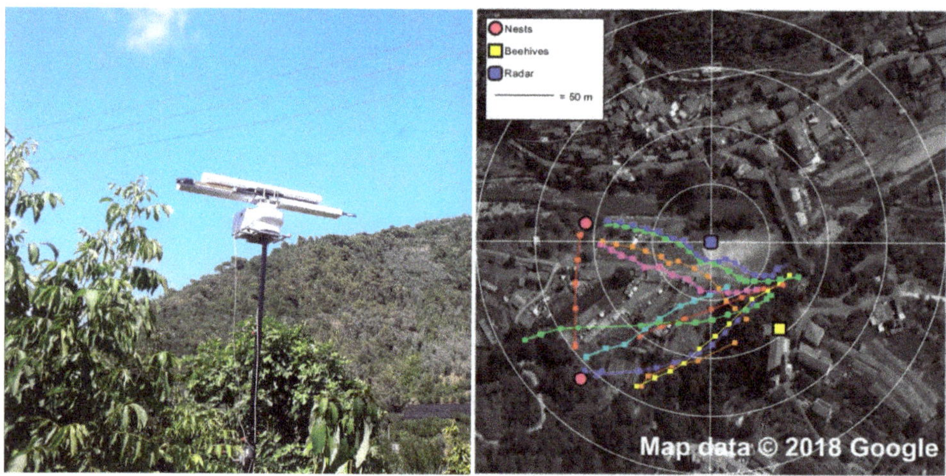

Figure 10. Harmonic radar able to follow the flight of hornets equipped with a passive transponder (left). On the map, the tracks obtained allowed to follow the path of the hornets and locate the nests (right).

5.2. Biological Control

Identifying organisms capable of parasitizing hornets could allow the selection of potential control agents, always remembering that they could also be transmitted to native species. At the moment, there are no known effective enemies or adversities in Europe that could be used for the biological control of *V. v. nigrithorax*. The main limitation for a biological control program is the lack of knowledge of the biology and ecology of *V. v. nigrithorax* in both the native and invaded territory.

In Asia *V. v. nigrithorax* is parasitized by *Bareogonalos jezoensis* (Yamane, 1973) (Hymenoptera Trigonalidae), but its use in biological control cannot be considered as a fighting agent in Europe since it would also parasitize other species of wasps or other insects. Before any use of parasites or other exotic biological agents, preventive and rigorous verifications are always necessary to rule out any effects on other native species [73].

In France, larval forms of the *Conops vesicularis* (Diptera Conopidae) were found, inside the abdomen of some individuals of *V. v. nigrithorax*, which resulted in their death [57,74]. Spradbery [75] reports that adults of *C. vesicularis* can wait at the entrance of the nest for homing workers, attack them and oviposit their eggs. So individuals of *V. v. nigrithorax* may be parasitized by this species in the environment, during foraging activity, or near the nest. However, the effectiveness of *C. vesicularis* as a biological agent for controlling *V. v. nigrithorax* populations appears limited.

Another potential parasite of *V. v. nigrithorax* has been confirmed by Villemant et al. [76], with the discovery of *Pheromermis vesparum* (Nematoda Mermithidae). This parasite was found in adult specimens of *V. v. nigrithorax* in France on two occasions, in November 2012 at Dompierre-sur-Besbre, and in January 2013 in Issigeac. However, even in this case, its effectiveness as biological control agent appears limited, as they are the only two cases of nematodes found on *V. v. nigrithorax* specimens throughout Europe.

In France, an entomopathogenic fungus (*Beauveria bassiana*) has been described to infect the common wasp *Vespula vulgaris* [77]. French researchers are studying if it is therefore likely to infect other hornets such *V. v. nigrithorax*. Poidatz et al. [78] describe *Metarhizium robertsii* as potential biological control agents against the invasive hornet *V. v. nigrithorax*.

Some species of mammals (as *Meles meles*) and birds (as *Garrulus glandarius*, *Merops apiaster*, *Parus major*, *Pica pica*, *Sitta europaea*, and *Gallus gallus domesticus*) can prey upon *V. v. nigrithorax*, but the predatory activity carried out by these animals is essentially sporadic and not enough to limit the population of the hornet; *Pernis apivorus* was also reported to exploit active *V. v. nigrithorax* nests [38,79].

6. Conclusions

The alien species *V. v. nigrithorax*, since its accidental introduction in France, has now successfully colonized several European countries. In the areas in which it has settled, it has become clear that it has a negative effect on *A. mellifera*, documented by the numerous losses of beehives reported by beekeepers. Hornets concentrate their predation activity on the honey bee colonies as they provide an abundant and continuous source of food. *V. v. nigrithorax*, however, hunts numerous other insects present in the environment and, among them, wild bees, attacking them while they are on the flowers intent on collecting nectar and pollen. This activity removes from the environment insects that play a very important ecosystem role. In fact, by visiting cultivated and spontaneous plants, wild pollinators guarantee not only the production of seeds and fruits of economic interest, but also the biodiversity of spontaneous plants.

Recent studies have shown that in Europe, and on other continents, pollinating insects and honey bees are in decline due to a combination of multiple factors. There is no doubt that the accidental introduction of the invasive predator *V. v. nigrithorax* can aggravate the situation. In fact, the Asian yellow-legged hornet has widely contributed to the decrease in the colonies of honey bees, weakening them to such an extent that they collapse and are more susceptible to parasites, viruses, and fungi.

Given the fundamental role of natural pollinators, studies are currently under way to ascertain the real effects of *V. v. nigrithorax* on wild bee populations and, more generally, on the environment in the newly introduced areas.

The introduction of an invasive exotic species, as well as causing damage to the ecosystem and biodiversity, can also generate a great deal of damage to the economy. *V. v. nigrithorax* has a negative impact, particularly on agriculture. The disappearance of beehives leads to the loss of honey bee products, putting the beekeeping industry in the position to be out of the market due to lack of production and/or rising of production costs. This type of economic impact is currently the most studied and can be easily expressed in monetary values.

The reduction in production yields of crops, as a result of the reduction of pollinators in general, is not quantifiable at the moment.

Author Contributions: Conceptualization, D.L. and A.M.; validation, S.L. and L.C.; literature investigation, D.L., A.M. and M.P.; data curation, D.L., S.L. and L.C.; writing—Original draft preparation, D.L.; writing—Review and editing, D.L. and S.L.; visualization, D.L. and L.C.; supervision, M.P.; project administration, S.L.; funding acquisition, M.P. All authors have read and agreed to the published version of the manuscript.

Funding: The paper was published thanks to University of Turin funds but should be considered as an After-LIFE product of the LIFE14 NAT/IT/001128 STOPVESPA funded by the European Commission.

Conflicts of Interest: The authors declare no conflict of interest, the absence of any personal circumstances or interest that may be perceived as inappropriately influencing the representation or interpretation of reported research results. The funder had no role in the design of the study; in the collection, analyses, or interpretation of data; in the writing of the manuscript, or in the decision to publish the results.

References

1. Atkinson, I.A.E. Introductions of wildlife as a cause of species extinctions. *Wildl. Biol.* **1996**, *2*, 135–141. [CrossRef]

2. Gandhi, K.J.K.; Herms, D.A. Direct and indirect effects of alien insect herbivores on ecological processes and interactions in forests of eastern North America. *Biol. Invasions* **2010**, *12*, 389–405. [CrossRef]

3. Lever, C. *Silent Summer: The State of Wildlife in Britain and Ireland—Vertebrate Animal Introductions*; Maclean, N., Ed.; Cambridge University Press: Cambridge, UK, 2010; Volume 4, pp. 36–52.

4. Van der Vecht, J. The Vespinae of the Indo Malayan and Papuan areas (Hymenoptera, Vespidae). *Zool. Verh.* **1957**, *34*, 1–83.

5. Van der Vecht, J. Notes on Oriental Vespinae, including some species from China and Japan (Hymenoptera, Vespidae). *Zool. Meded.* **1959**, *36*, 205–232.

6. Archer, M.E. Taxonomy, distribution and nesting biology of the *Vespa bicolor* group (Hym., Vespinae). *Entomol. Mon. Mag.* **1994**, *130*, 149–158.

7. Perrard, A.; Arca, M.; Rome, Q.; Muller, F.; Tan, J.; Bista, S.; Ho, H.N.; Baudoin, R.; Baylac, M.; Silvain, J.F.; et al. Geographic variation of melanisation patterns in a hornet species: Genetic differences, climatic pressures or aposematic constraints? *PLoS ONE* **2014**, *9*, e94162. [CrossRef]

8. Archer, M.E. *Vespine Wasps of the World. Behaviour, Ecology and Taxonomy of the Vespinae*; Monograph Series; Siri Scientific Press: Manchester, UK, 2012; Volume 4, pp. 1–352. ISBN 978-0-9567795-7-1.

9. Choi, M.B.; Martin, S.J.; Lee, J.W. Distribution, spread, and impact of the invasive hornet *Vespa velutina* in South Korea. *J. Asia Pac. Entomol.* **2012**, *15*, 473–477. [CrossRef]

10. Sakay, Y.; Takahashi, J. Discovery of a worker of *Vespa velutina* (Hymenoptera: Vespidae) from Tsushima Island, Japan. *Jpn. J. Entomol.* **2014**, *17*, 32–36.

11. Ueno, T. Establishment of the invasive hornet *Vespa velutina* (Hymenoptera: Vespidae) in Japan. *Int. J. Chem. Environ. Biol. Sci.* **2014**, *2*, 220–222.

12. Minoshima, Y.N.; Yamane, S.; Ueno, T. An invasive alien hornet, *Vespa velutina nigrithorax* du Buysson (Hymenoptera, Vespidae), found in Kitakyushu, Kyushu Island: A first record of the species from mainland Japan. *Jpn. J. Syst. Entmol.* **2015**, *21*, 259–261.

13. Takahashi, J.; Okuyama, H.; Kiyoshi, T.; Takeuchi, T.; Martin, S.J. Origins of *Vespa velutina* hornets that recently invaded Iki Island, Japan and Jersey Island, UK. *Mitochondrial DNA* **2018**, 1–6. [CrossRef] [PubMed]

14. Haxaire, J.; Bouguet, J.P.; Tamisier, J.P. *Vespa velutina* Lepeletier, 1836, une redoutable nouveauté pour la faune de France (Hym., Vespidae). *Bull. Soc. Entomol. Fr.* **2006**, *111*, 194.

15. Villemant, C.; Haxaire, J.; Streito, J.L. Premier bilan de l'invasion de *Vespa velutina* Lepeletier en France (Hymenoptera, Vespidae). *Bull. Soc. Entomol. Fr.* **2006**, *111*, 535–538.

16. Castro, L.; Pagola-Carte, S. *Vespa velutina* Lepeletier, 1836 (Hymenoptera: Vespidae) recolectada en la Península Ibérica. *Heteropterus Rev. Entomol.* **2010**, *10*, 193–196.

17. Lopéz, S.; Gonzalez, M.; Goldarazena, A. *Vespa velutina* Lepeletier, 1836 Hymenoptera: Vespidae): First records in Iberian Peninsula. *Bull. OEPP/EPPO Bull.* **2011**, *41*, 439–441. [CrossRef]

18. Rodríguez-Flores, M.S.; Seijo-Rodríguez, A.; Escuredo, O.; Seijo-Coello, M.C. Spreading of *Vespa velutina* in northwestern Spain: Influence of elevation and meteorological factors and effect of bait trapping on target and non-target living organisms. *J. Pest Sci.* **2019**, *92*, 557–565. [CrossRef]

19. Pujade-Villar, J.; Torrel, A.; Rojo, M. Confrimada la presència a CAtalunyad'una vespa originària d'Asia molt perillosa per als ruscs. *Butll. Inst. Cat. Hist. Nat.* **2012**, *77*, 173–176.

20. Leza, M.; Miranda, M.A.; Colomar, V. First detection of *Vespa velutina nigrithorax* (Hymenoptera: Vespidae) in Balearic Islands (Western Mediterranean): A challenging study case. *Biol. Invasions* **2018**, 1–7. [CrossRef]

21. Leza, M.; Herrera, C.; Marques, A.; Roca, P.; Sastre Serra, J.; Pons, D.G. The impact of the invasive species *Vespa velutina* on honeybees: A new approach based on oxidative stress. *Sci. Total Environ.* **2019**, *689*, 709–715. [CrossRef]

22. Grosso-Silva, J.M.; Maia, M. *Vespa velutina* Lepeletier, 1836 (Hymenoptera, Vespidae), new species for Portugal. *AEGA* **2012**, *6*, 53–54.

23. Rome, Q.; Muller, F.; Villemant, C. Expansion en 2011 de *Vespa velutina* Lepeletier en Europe (Hym., Vespidae). *Bull. Soc. Entomol. Fr.* **2012**, *117*, 114.

24. Demichelis, S.; Manino, A.; Minuto, G.; Mariotti, M.; Porporato, M. Social wasp trapping in north west Italy: Comparison of different bait-traps and first record of *Vespa velutina* Lepeletier (Hymenoptera: Vespidae). *Bull. Insectol.* **2014**, *67*, 307–317. [CrossRef]

25. Porporato, M.; Manino, A.; Laurino, D.; Demichelis, S. *Vespa velutina* Lepeletier (Hymenoptera Vespidae): A first assessment two years after its arrival in Italy. *Redia* **2014**, *97*, 189–194.

26. Bertolino, S.; Lioy, S.; Laurino, D.; Manino, A.; Porporato, M. Spread of the invasive yellow legged hornet *Vespa velutina* (Hymenoptera: Vespidae) in Italy. *Appl. Entomol. Zool.* **2016**, *51*, 589–597. [CrossRef]

27. Lioy, S.; Manino, A.; Porporato, M.; Laurino, D.; Romano, A.; Capello, M.; Bertolino, S. Establishing surveillance areas for tackling the invasion of *Vespa velutina* in outbreaks and over the border of its expanding range. *Neobiota* **2019**, *46*, 51–69. [CrossRef]

28. Witt, R. Erstfund eines Nestes der Asiatischen Hornisse *Vespa velutina* Lepeletier, 1838 in Deutschland und Details zum Nestbau (Himonoptera, Vespinae). *Ampulex* **2015**, *7*, 42–53.

29. UK National Bee Unit. 2016. Available online: http://nationalbeeunit.com/ (accessed on 9 April 2019).

30. Budge, G.E.; Hodgetts, J.; Jones, E.P.; Ostojà-Starzewski, J.C.; Tomkies, V.; Semmence, N.; Brown, M.; Stainton, K. The invasion provenance and diversity of *Vespa velutina* Lepeletier (Hymenoptera: Vespidae) in Great Britain. *PLoS ONE* **2017**, *12*, 1–12. [CrossRef]

31. Villemant, C.; Barbet-Massin, M.; Perrard, A.; Muller, F.; Gargominy, O.; Jiguet, F.; Rome, Q. Predicting the invasion risk by the alien bee-hawking Yellow-legged hornet *Vespa velutina nigrithorax* across Europe and other continents with niche models. *Biol. Conserv. J.* **2011**, *144*, 2142–2150. [CrossRef]

32. Barbet-Massin, M.; Rome, Q.; Muller, F.; Perrard, A.; Villemant, C. Climate change increases the risk of invasion by the Yellow-legged hornet. *Biol. Conserv.* **2013**, *157*, 4–10. [CrossRef]

33. Kishi, S.; Goka, K. Review of the invasive yellow-legged hornet, *Vespa velutina nigrithorax* (Hymenoptera: Vespidae), in Japan and its possible chemical control. *Appl. Entomol. Zool.* **2017**, *52*, 361–368. [CrossRef]

34. Rome, Q.; Muller, F.J.; Touret-Alby, A.; Darrouzet, E.; Perrard, A.; Villemant, C. Caste differentiation and seasonal changes in *Vespa velutina* (Hym.: Vespidae) colonies in its introduced range. *J. Appl. Entomol.* **2015**, *139*, 771–782. [CrossRef]

35. Monceau, K.; Bonnard, O.; Thiery, D. *Vespa velutina*: A new invasive predator of honeybees in Europe. *J. Pest Sci.* **2014**, *87*, 1–16. [CrossRef]

36. Wang, Z.; Qu, Y.; Dong, S.; Wen, P.; Li, J.; Tan, K.; Menzel, R. Honey bees modulate their olfactory learning in the presence of hornet predators and alarm component. *PLoS ONE* **2016**, *11*, 1–12. [CrossRef]

37. Couto, A.; Monceau, K.; Bonnard, O.; Thiéry, D.; Sandoz, J.C. Olfactory attraction of the hornet *Vespa velutina* to honeybee colony odors and pheromones. *PLoS ONE* **2014**, *9*, 1–19. [CrossRef]

38. Laurino, D.; Porporato, M. *Vespa Velutina—Conoscerla e Prepararsi ad Affrontare il Pericolo*, 1st ed.; Edizioni Montaonda: San Godenzo, Italy, 2017; p. 67. ISBN 9-78898-186228.

39. Wang, Z.W.; Chen, G.; Tan, K. Both olfactory and visual cues promote the hornet *Vespa velutina* to locate its honeybee prey Apis Cerana. *Insectes Soc.* **2014**, *61*, 67–70. [CrossRef]

40. Turchi, L.; Derijard, B. Options of the biological and physical control of *Vespa velutina nigrithorax* (Hym.: Vespidae) in Europe: A review. *J. Appl. Entomol.* **2018**, *1*, 10. [CrossRef]

41. Couto, A.; Lapeyre, B.; Thiery, D.; Sandoz, J.C. Olfactory pathway of the hornet *Vespa velutina*: New insights into the evolution of the hymenopteran antennal lobe. *J. Comp. Neurol.* **2016**, *524*, 2335–2359. [CrossRef]

42. Wen, P.; Cheng, Y.N.; Dong, S.H.; Wang, Z.W.; Tan, K.; Nieh, J.C. The sex pheromone of a globally invasive honey bee predator, the Asian eusocial hornet, *Vespa velutina*. *Sci. Rep.* **2017**, *7*, 12956. [CrossRef]

43. Cheng, Y.; Wen, P.; Dong, S.; Tan, K.; Nieh, J.C. Poison and alarm: The Asian hornet *Vespa velutina* uses sting venom volatiles as alarm pheromone. *J. Exp. Biol.* **2017**, *220*, 645–651. [CrossRef]

44. Fournier, A.; Barbet-Massin, M.; Rome, Q.; Courchamp, F. Predicting species distribution combining multi-scale drivers. *Glob. Ecol. Conserv.* **2017**, *12*, 215–226. [CrossRef]

45. Robinet, C.; Suppo, C.; Darrouzet, E. Rapid spread of the invasive yellow-legged hornet in France: The role of human-mediated dispersal and the effects of control measures. *J. Appl. Ecol.* **2017**, *54*, 205–215. [CrossRef]

46. Barbet-Massin, M.; Rome, Q.; Villemant, C.; Courchamp, F. Can species distribution models really predict the expansion of invasive species? *PLoS ONE* **2018**, *13*, e0193085. [CrossRef] [PubMed]

47. Beggs, J.R.; Brockerhoff, E.G.; Corley, J.C.; Kenis, M.; Masciocchi, M.; Muller, F.; Rome, Q.; Villemant, C. Ecological effects and management of invasive alien Vespidae. *BioControl* **2011**, *56*, 505–526. [CrossRef]

48. Monceau, K.; Maher, N.; Bonnard, O.; Thiery, D. Predation pressure dynamics study of the recently introduced honeybee killer *Vespa velutina*: Learning from the enemy. *Apidologie* **2013**, *44*, 209–221. [CrossRef]

49. Monceau, K.; Bonnard, O.; Moreau, J.; Thiéry, D. Spatial distribution of *Vespa velutina* individuals hunting at domestic honeybee hives: Heterogeneity at a local scale. *Insect Sci.* **2014**, *21*, 765–774. [CrossRef] [PubMed]

50. Moller, H. Lessons for invasion theory from social insects. *Biol. Conserv.* **1996**, *78*, 125–142. [CrossRef]

51. Martin, S.J. *The Asian hornet: Threats, Biology & Expansion*; The International Bee Research Association: England and Wales, 2017; p. 106. ISBN 9780860982814.

52. De Haro, L.; Labadie, M.; Chanseau, P.; Cabot, C.; Blanc-Brisset, I.; Penouil, F. Medical consequences of the Asian black hornet (*Vespa velutina*) invasion in Southwestern France. *Toxicon* **2010**, *55*, 650–652. [CrossRef]

53. Feàs Sanchez, X.; Charles, R.J. Notes on the Nest Architecture and Colony Composition in Winter of the Yellow-Legged Asian Hornet, *Vespa velutina* Lepeletier 1836 (Hym.: Vespidae), in Its Introduced Habitat in Galicia (NW Spain). *Insects* **2019**, *10*, 237. [CrossRef]

54. Liu, Z.R.; Chen, S.G.; Zhou, Y.; Xie, C.H.; Zhu, B.F.; Zhu, H.M.; Liu, S.P.; Wang, W.; Chen, H.Z.; Ji, Y.H. Deciphering the Venomic Transcriptome of Killer-Wasp *Vespa velutina*. *Sci. Rep.* **2015**, *5*, 9454. [CrossRef]

55. Tabar, A.I.; Chugo, S.; Joral, A.; Lizaso, M.T.; Lizarza, S.; Alvarez-Puebla, M.J.; Arroabarren, E.; Vela, C.; Lombardero, M. *Vespa velutina nigrithorax*: A new causative agent for anaphylaxis. *Clin. Transl. Allergy* **2015**, *5*, 43. [CrossRef]

56. Sumner, S.; Law, G.; Cini, A. Why we love bees and hate wasps. *Ecol. Entomol.* **2018**, *43*, 836–845. [CrossRef]

57. Requier, F.; Rome, Q.; Chiron, G.; Decante, D.; Marion, S.; Menard, M.; Muller, F.; Villemant, C.; Henry, M. Predation of the invasive Asian hornet affects foraging activity and survival probability of honey bees in Western Europe. *J. Pest Sci.* **2019**, *92*, 567–578. [CrossRef]

58. Monceau, K.; Thiery, D. *Vespa velutina*: Current situation and perspectives. *Atti Accad. Naz. Ital. Entomol.* **2016**, *64*, 137–142.

59. Helmer, S.H.; Kerbaol, A.; Aras, P.; Jumarie, C.; Boily, M. Effects of realistic doses of atrazine, metolachlor, and glyphosate on lipid peroxidation and diet-derived antioxidants in caged honey bees (*Apis mellifera*). *Environ. Sci. Pollut. Res.* **2015**, *22*, 8010–8021. [CrossRef]

60. Rome, Q.; Perrard, A.; Muller, F.; Villemant, C. Monitoring and control modalities of a honeybee predator, the Yellow-legged hornet *Vespa velutina nigrithorax* (Hymenoptera: Vespidae). *Aliens Invasive Spec. Bull.* **2011**, *31*, 7–15.

61. Ken, T.; Hepburn, H.R.; Radloff, S.E.; Yusheng, Y.; Yiqiu, L.; Danyin, Z.; Neumann, P. Heat-balling wasps by honeybees. *Naturwissenschaften* **2005**, *92*, 492–495. [CrossRef]

62. Genovesi, P.; Scalera, R.; Brunel, S.; Solarz, W.; Roy, D. *Towards an Early Warning and Information System for Invasive Alien Species (IAS) Threatening Biodiversity in Europe*; Tech. Report 5/2010; European Environment Agency: Copenhagen, Denmark, 2010; 52p. [CrossRef]

63. Rojas-Nossa, S.V.; Novoa, N.; Serrano, A.; Calviño-Cancela, M. Performance of baited traps used as control tools for the invasive hornet *Vespa velutina* and their impact on non-target insects. *Apidologie* **2018**, *49*, 872–885. [CrossRef]

64. Blot, J. Localisation et destrution des nids de frelons asiatiques. *Bull. Tech. Apic.* **2008**, *35*, 95–100.
65. Jeffs, K. Wild China BBC, Hornet Sequence between 42' and 45'. 2008. Available online: http://www.bbc.co.uk/iplayer/episode/b==brvjx/wild-china-2-shangrila;https://vimeo.com/7877242 (accessed on 15 October 2019).
66. Reynaud, B.; Guérin-Lassous, I. Design of a force-based controlled mobility on aerial vehicles for pest management. *Ad Hoc Netw.* **2016**, *53*, 41–52. [CrossRef]
67. Semmence, N. Asian Hornet Update from the National Bee Unit. *Incorporating the British Bee J. BBKA News*, November 2018; 270–272.
68. Kennedy, P.; Ford, S.M.; Poidatz, J.; Thiéry, D.; Osborne, J.L. Searching for nests of the invasive Asian hornet (*Vespa velutina*) using radio-telemetry. *Commun. Biol.* **2018**, *1*, 88. [CrossRef]
69. Bortolotti, L.; Cervo, R.; Felicioli, A.; Quaranta, M.; Salvetti, O.; Berton, A. Progetto Velutina: La ricerca italiana a caccia di soluzioni. *Atti Accad. Naz. Ital. Entomol.* **2016**, *64*, 143–149.
70. Milanesio, D.; Saccani, M.; Maggiora, R.; Laurino, D.; Porporato, M. Design of an harmonic radar for the tracking of the Asian yellow-legged hornet. *Ecol. Evol.* **2016**, *6*, 2170–2217. [CrossRef] [PubMed]
71. Milanesio, D.; Saccani, M.; Maggiora, R.; Laurino, D.; Porporato, M. Recent upgrades of the harmonic radar for the tracking of the Asian yellow-legged hornet. *Ecol. Evol.* **2017**, *7*, 4599–4606. [CrossRef] [PubMed]
72. Maggiora, R.; Saccani, M.; Milanesio, D.; Porporato, M. An Innovative Harmonic Radar to Track Flying Insects: The Case of *Vespa velutina*. *Sci. Rep.* **2019**, *9*, 1–10. [CrossRef]
73. Yamane, S. New taxa of the genius *Bareogonalos* from Asia with further information on the tribe Nomadinini (Hymenoptera, Trigonalidae). *Halteres* **2014**, *5*, 17–31.
74. Darrouzet, E.; Gérvar, J.; Dupont, S. A scientific note about a parasitoid that can parasitize the yellow-legged hornet, *Vespa velutina nigrithorax*, in Europe. *Apidologie* **2014**, *46*, 130–132. [CrossRef]
75. Spradbery, J.P. Chap 6: Feeding and Foraging. In *Wasps. An Account of the Biology and Natural History of Solitary and Social Wasps*; Sidgwick & Jackson: London, UK, 1973; pp. 128–157.
76. Villemant, C.; Zuccon, D.; Rome, Q.; Muller, F.; Poinar, G.O., Jr.; Justine, J.L. Des parasites peuvent-ils stopper l'invasion? Des nématodes mermithidés parasitent le frelon asiatique à pattes jaunes en France. *Peer J.* **2015**, *947*. [CrossRef]
77. Harris, R.; Harcourt, S.J.; Glare, T.R.; Rose, E.A.F.; Nelson, T.J. Susceptibility of *Vespula vulgaris* (Hymenoptera: Vespidae) to Generalist Entomopathogenic Fungi and their potential for wasp control. *J. Invertebr. Pathol.* **2000**, *75*, 251–258. [CrossRef]
78. Poidatz, J.; López Plantey, R.; Thiéry, D. Indigenous strains of *Beauveria* and *Metharizium* as potential biological control agents against the invasive hornet *Vespa velutina*. *J. Invertebr. Pathol.* **2018**, *153*, 180–185. [CrossRef]
79. Macià, F.X.; Menchetti, M.; Corbella, C.; Grajera, J.; Vila, R. Exploitation of the invasive Asian Hornet *Vespa velutina* by the European Honey Buzzard *Pernis apivorus*. *Bird Study* **2019**, *66*, 425–429. [CrossRef]

 *diversity*

Review

Diagnosis of Varroa Mite (*Varroa destructor*) and Sustainable Control in Honey Bee (*Apis mellifera*) Colonies—A Review

Aleš Gregorc [1],* and Blair Sampson [2]

[1] Faculty of Agriculture and Life Sciences, University of Maribor, Pivola 10, 2311 Hoče, Slovenia
[2] Thad Cochran Southern Horticultural Research Laboratory, USDA, ARS, Poplarville, MS 39470, USA; Blair.Sampson@ars.usda.gov
* Correspondence: ales.gregorc@um.si

Received: 15 November 2019; Accepted: 12 December 2019; Published: 16 December 2019

Abstract: Determining varroa mite infestation levels in honey bee colonies and the proper method and time to perform a diagnosis are important for efficient mite control. Performing a powdered sugar shake or counting mites that drop from combs and bees onto a hive bottom board are two reliable methods for sampling varroa mite to evaluate the efficacy of an acaricide treatment. This overview summarizes studies that examine the efficacy of organic acids and essential oils, mite monitoring, and brood interruption for integrated varroa mite control in organic beekeeping.

Keywords: varroa mite detection; diagnosis; infestation; mortality; control; organic treatment

---

1. Introduction

Varroa mites, *Varroa destructor* Anderson and Truemann, are the single most devastating global pests of both adult and immature honey bees, *Apis mellifera* L. Feeding by mites reduces adult bee body weight, life span, and immunity to pathogens [1–4], Adult female varroa mites are physically large (~1.5–2.0 mm wide) parasitic mites that disperse through phoresy by attaching themselves to worker bees and drones. Before worker bees seal comb cells with wax, varroa mites move into brood cells occupied by mature bee larvae where they will eventually feed on the fat bodies and hemolymph of host pupae [5,6]. If mite populations develop undetected, untreated and infested honey bee colonies usually collapse within a year [3]. Although there are multiple factors contributing to colony mortality, in many cases, the most important are varroa parasitism [4] together with the transmission of an extensive set of honey bee viruses [5]. The chemical control of varroa mites presents a paradox. Toxic acaricides must be used to kill mites, but they can also have toxic side-effects on bees and whole colonies. Thus, acaricides must be applied at their lowest effective dose to minimize a build-up of chemical residues and their by-products in bees, honey, and wax [6], as well as to slow the evolution of acaricide resistance [7]. Improved monitoring for mites and the application of less-persistent, low-residual acaricides could help reduce the amount of toxic active products applied each season. Therefore, we review research regarding improvements in mite monitoring and the efficacy of alternative organic acaricides.

Evaluation of Colonies Infestation with Varroa Mites

Researchers and beekeepers employ various standardized or incidental methods to monitor varroa mites infesting adult honey bees and brood [8]. These methods include counting mites on traps and on such colony contents as hive debris on the bottom boards, brood, as well as the bodies of captured worker bees. Sampling such hive contents is crucial for collecting data for optimally timing acaricide applications and effectively controlling mites [9].

Currently, simple field and laboratory diagnostic procedures are available for assessing overall varroa mite infestation within entire honey bee colonies. Beekeepers have been evaluating for decades the efficacy of an acaricide treatment by counting dead mites that drop from brood frames and bees onto the hive bottom board [10,11]. The number of these fallen mites found within a hive's debris correlates well with living mite populations infesting the colony above [12]. In fact, researchers studied the correlation between the natural mite mortality and the total number of mites in honey bee colonies. They found that varroa mite populations in colonies with brood can be approximated by multiplying the daily varroa mite mortality detected on hive bottom boards by 20–40 [13]. After performing mite counts and once infestations of varroa have reached or exceeded economic threshold level, e.g., at 3 mites/100 bees in the colony, control is warranted [13]. However, treatment threshold levels will vary according to colony size, location, management and other stress factors [14]. Therefore, keeping data records of natural mite fall levels within a colony is useful for determining (1) the level of mite infestation before and after acaricidal application and (2) seasonal trends in the size of natural mite populations [15]. Bottom board counts are especially useful for estimating total varroa mites per colony and daily mite mortality, even for very small mite populations [16], because varroa mite populations below 100 mites per colony cannot be detected by examining samples of adult bees or brood [17]. However, mite sampling protocols that rely on bottom board mite counts are time consuming, more costly, and may take several days or weeks to establish a relevant colony infestation level. Varroa mite counts may also vary with levels of mite removal by bees (i.e., hygienic behavior) [18].

A simple non-destructive method for sampling varroa mites living on the bodies of adult honey bees is the "sugar shake". This technique simply involves applying powdered sugar to the bodies of living bees. The powdered sugar will quickly clog the tarsal pads of varroa mites, they lose adhesion, and become permanently dislodged from host bees [19]. The sugar shake method removes 77% to 91% of mites for counting [20,21]. Even in highly infested colonies, 82% recovery of varroa mites is possible using the sugar shake method [22]. The size of mite populations in an entire colony can be extrapolated from counts of mites dislodged from adult bees taken from a single brood frame. Higher precision in estimates of colony infestation simply requires counting dislodged mites from a greater number of captured bees [8]. Hence, the "sugar shake" procedure is a simple and reliable method that beekeepers can use to evaluate the number of varroa mites on adult worker bees, drones, and even throughout an entire colony [22,23]. Sugar shake is also a useful sampling protocol when evaluating different mite controls [24]. There are a few caveats to consider when using the sugar shake method. The efficiency of the powder shake method to extract varroa mites from adult host bees can vary with fluctuations in honey bee population size and environmental conditions. Average effectiveness of the sugar shake method to remove varroa mites from adult honey bees in a hot, humid environment (i.e., 32 °C and 76% RH) is ~66%, a rate below the 94% obtained in drier, cooler conditions (i.e., 26 °C and 71% RH) [25]. Sugar quality is also important for dislodging varroa, with some sugar mixtures outperforming others. For instance, a fine dusting of pure powdered sugar dislodges 70–80% of varroa mites, which greatly exceeds the 50% varroa drop achieved using a mix of powdered sugar and corn starch. Thus, sugar quality combined with a humid environment appears to influence the efficacy of sugar shakes to remove mites [19]. Perhaps, a higher temperature and humidity causes grains to clump, thereby increasing the sugar's coarseness and reducing rates of tarsal clogging. Generally, dusting honey bees with powdered sugar is considered a safe mechanical method for diagnosing colony infestations and in some cases, can be used to reduce varroa mite infestation in field colonies [24].

A similar approach to removing mites from host bees is to use washing kits, often homemade, with water or alcohol (70%) serving as both the washing and collection fluid [26]. Accurate estimates of the number of varroa mites on adults using either the sugar shake method, water wash, or alcohol wash can be performed in colonies without brood for research or practical beekeeping purposes. When using an alcohol wash, approximately 300 bees are sacrificed without any noticeable effect on colony strength or overall health and productivity. However, alcohol washes will kill both mites and the small number of bees selected for sampling. In fact, accurate assessments of colony infestation obtained with this diagnostic test are crucial to ensuring colony survival and productivity [27]. Based on an established number of varroa mites obtained by sugar shake or alcohol wash, applications of miticides or other treatments may be warranted. Therefore, the efficacy of previous control treatments can be assessed using both methods. Varroa mite levels between 3 percent and 5 percent are tolerable for beekeepers. When mite levels are above 5 percent, immediate varroa mite control needs to be performed [28,29]. A similar wide range of economic threshold for varroa mite populations in colonies was established for beekeeping conditions in the United States at levels between 5 and 12 percent and most beekeepers prefer to use the 5 percent [13]. Precise estimation of varroa mite population size is established by surveying mite loads on adult and pupal bees as well as through counts of dead varroa mites within a colony. The population dynamics of varroa mites mainly depend on the reproductive success of varroa females in worker brood cells [30,31]. Furthermore, mite population size may be estimated by a combination of adult bee and brood samples and compared to the number of mites killed by chemical treatment [32].

2. Varroa Mite Control with Organic Substances

Proper timing of control is best accomplished by regularly monitoring mite levels inside colonies and comparing varroa mite population trends. Organic acids derived mostly from active components of plant and essential oils are widely used for varroa mite control in conventional and organic honeybee colonies [11,33]. Thymol is one such component, which is a natural constituent of thyme (*Thymus vulgaris*) and a volatile monoterpenoid with acaricidal activity against varroa [28]. The efficacy and environmental safety of several commercial products containing thymol have been evaluated [11, 29,30]. It has been established that thymol residues may accumulate in a hive, but they soon dissipate to a natural hive level at or below 1.1 mg thymol/kg honey. Thus, thymol administered to a hive's population does not alter the taste or nutritional quality of harvested honey [31]. Two popular thymol-based acaricides are available for beekeeping. They include ApiGuard® and Thymovar®. ApiGuard is a registered thymol-based fumigant that in its gel form controls high percentages of varroa mites in honeybee colonies. In broodless colonies, ApiGuard can kill 54 percentage points more varroa mites (77% mortality) than does natural mortality alone (23% natural mortality), [14]. Similarly, Fasbinder et al. [33] attained 64% mortality with ApiGuard. The most effective applications of ApiGuard occur in autumn, a time when this acaricide has little to no negative impact on the growth of early spring bee populations [34]. However, like many other acaricides, the effectiveness of ApiGuard treatments varies according to environmental conditions. Thymovar (Andrematt Biocontrol AG) is another thymol-based product formulated as 15 g of thymol impregnated onto a cellulose wafer. This product is typically applied after the last honey harvest in the beekeeping season. In one of our previous experiments, we found that Thymovar kills 54–66% of varroa mites [32] and in another experiment, [35] attained a mean percentage efficacy of 85%.

Other non-persistent organic acids with high acaricidal activity include formic acid (FA), a common defensive compound sprayed by certain arthropods, and oxalic acid (OA), a minor constituent of honey and a plant substance that, in a normal field context, repels or sickens a plant's insect antagonists [30,36]. Another organic acaricide with high activity, derived from hop beta plant acids, is HopGuard®. It is particularly effective at killing varroa mites in colonies with both open and sealed brood. The recommended timing of HopGuard is during the months of June and October in colonies with large brood areas [37]. HopGuard efficacy can vary seasonally and among colonies, but it often induces >80% varroa mite mortality in brood-right colonies. HopGuard works quickly and kills 65% of varroa mites within 24 h of application [38].

FA treatment alone kills 43% of mites on the bodies of adult worker bees, but is less effective in autumn when colonies receive double fumigation with 65% FA [39]. Improved formulations of FA (50 mL and 100 mL at 85%) impregnating thin plates of soft treefiber induces 100% mortality of mother mites with 90% of honey bee brood surviving. Higher bee brood loss did occur with a treatment of 100 mL FA [40]. Absorbent cardboard plates (Illertissen mite plates) soaked with 20 mL of 65% formic acid induces 94% varroa mite mortality in Schwarzwald (Black Forest) colonies [41]. Higher varroa mite control efficiencies of 95–97% using FA impregnated soft fiber plates (Pavatex) depended on both concentration and duration [42]. Formic acid as a volatile organic compound dissipates quickly after treatment and hence, has a very short period of residual activity [43]. At higher temperatures, FA evaporation increases, and if hive ventilation is impeded by bees covering vents with propolis, mite mortality may fall, while brood mortality increases [40].

Oxalic acid (OA) is acutely toxic to varroa mites. Due to this acaricide's high efficacy and low risk of hive contamination, OA has been extensively used by itself and in combination with other mite controls [30,36]. Application typically involves trickling OA in sugar water solution between the frames of a honey bee hive. The recommended application rate per hive is 50 mL of OA solution. Various formulations of OA that have been tested include 3.4% OA dihydrate in 47.6% aqueous sucrose (w/w), 3.7% OA in 27.1% aqueous sucrose, and 3.0% OA in 32.0% aqueous sucrose [44,45]. Oxalic acid can kill 33–69% of varroa mites within the first two days following application, particularly in summer and autumn. From 3 to 11 days post-treatment, OA residual activity drops off significantly to 14–23% after the first 3 and 4 days and to almost 0% after 9 days [46]. Higher concentrations of OA in sugar water does increase varroa mite control but may pose a greater toxicity risk to bees. In fact, honey bees are susceptible to the toxic effects of both conventional and organic acaricides, including OA even at low concentrations [45]. In toxicity tests using caged bees, the highest dosage of OA acaricide is approximately 250 µg per bee, which is a tolerance level 2.5 X higher than previously reported [47]. However, OA is less toxic to honey bees than is the conventional acaricide Apistan®. We suspect that individual bees are killed by direct dermal exposure to OA in both laboratory tests and colony treatments. While some bees are exposed to lethal doses of OA when using the trickling method within colonies, queens have a very low probability of exposure, yet mortality within any caste of honey bee could affect overall colony productivity, unless steps are taken to limit brood and adult exposure during initial acaricide applications [24].

Single applications of OA are the least harmful to bees, but may be insufficient for adequate varroa control. Brødsgaard [48] reported only a 24% efficacy after administering a single OA treatment in the spring using the trickling method. Repeated applications of OA during specific periods of colony development substantially improve varroa mite control. Three consecutive OA treatments in colonies maintained in a temperate environment with brood present reduced varroa mite populations by 39–52%. It seems that colonies with a threshold level less than 1 dead mite found on the hive bottom board per day by the end of July should receive three OA treatments, which will reduce the mite population by approximately 40% [49]. Reducing the mite population ensures a colony's normal development and winter survival. However, OA efficacy rises considerably to 98–99% when colonies lack brood [49,50]. Other researchers reported similar results with brood-right and broodless colonies in different climactic regimes [32,42,44,48]. Additionally, Nanetti et al, [50] and Imdorf et al. [51] eliminated 97.3–99.5% of mites with OA in broodless colonies, whereas Mutinelli et al. [52] achieved 95% mite control with three applications of 5% OA in colonies with capped brood. When using the trickling method in colonies with capped brood during the winter, the cumulative efficacy of four oxalic acid applications ranged from 90% to 100% with a mean value of 98% [53]. In the study by Toufailia and Ratnieks [53], efficacy values may have been overestimated because the experiment included a final Apistan application, which sometimes is less than 100% effective.

As alluded to previously, acaricidal efficacy of repeated OA applications vary during specific periods of colony development. When compared with a relatively low efficacy during the brood season, oxalic acid application results in higher cumulative varroa mite mortalities in overwintering colonies lacking brood. When capped brood is present, ~70% of the phoretic (adult) varroa are killed within two days of OA treatment [46,54]. Three summer applications of 2.9% OA in 31.9% sugar water (w/w) controls 30–52% of varroa mites [50], a range within that discovered by Toufailia et al. [55] (20–93% mite mortality) using similar OA formulations. In brood-right colonies kept in the Mediterranean area (Sardinia, Italy) with brood present, three OA applications in late autumn resulted in 65% varroa mite mortality [54,56]. Higher mite mortalities occur in winter for OA-treated colonies containing brood and in summer, for colonies receiving repeated OA applications. Further research is necessary to assess bee safety and acaricidal effectiveness of oxalic acid in honey bee colonies overwintering in warm temperate climates. A combination of four successive oxalic acid applications during warm temperate winter conditions ensures a substantial reduction in varroa mite populations before colony growth accelerates in spring. The high correlation between the natural mite-mortality and the mite mortality induced after multiple OA treatments during the summer indicates that the continual monitoring of mite drops within bee colonies is important [46].

As we have seen, greater mite control during cooler months indicates that improved varroa mite control is achievable by integrating OA control with a colony's broodless period. Broodless periods may not naturally occur in colonies maintained in warm-temperate to tropical climates where honey bees may reproduce year-round. Modelling seasonal mite population dynamics is needed, however, to determine when to artificially or naturally break brood rearing before applying acaricides like OA [57]. However, in warmer regions where bees tend brood in winter, colonies can be rendered artificially broodless by interrupting brood production through brood removal, particularly drone brood, or queen exclusion [30,58,59]. Coupling screened bottom boards with drone brood removal or queen caging can further interfere with mite reproduction and development [17,29,37]. The removal of brood and the queen also concentrates OA treatment on the phoretic form of varroa mites, reducing the likelihood of OA harming brood and reproductives, and hence, a colony's future productivity. Queen caging combined with a single application of an OA or thymol-based acaricide can increase the efficacy of varroa mite control to 96–97% [60,61]. The temporary caging of the queen, combined with the removal of capped or un-capped brood, has resulted in >93% varroa mite control [62]. The vaporisation OA crystals in colonies without brood is another varroa mites control method. OA crystals are placed into the small cup of the electrically heated applicator inserted under the bee cluster on the hive bottom. Vaporisation 2.25 g of OA per broodless colony induced 97.6% varroa mites mortality with 98% colonies overwinter survival [55]. Experiments were also conducted to test the efficacy of 10 g of OA mixed with 20 mL of glycerin impregnated cellulose strips. The average efficacies in three experimental locations in Argentina were in the rage of 93% to 94% [63].

Colonies with worker bees displaying a propensity to remove varroa mites along with sick brood (bee hygiene) may also be more safely treated with OA. A combination of mite removal by bees and OA applications could be a highly effective method for varroa management, particularly in organic operations where the use of conventional acaricides is prohibited. Comparing all three organic acaricides applied against varroa in colonies containing capped brood, Thymovar is more effective than FA or OA on a per application basis. However, two OA sublimations applied in autumn are as effective, or more effective than Thymol, killing 82% of varroa mites [37]. In contrast, two formic acid treatments did not reach the 85% Thymovar efficacy achieved by [36].

3. Integrated Varroa Mite Management

The main objective of integrated varroa management is to reduce or delay the use of chemical treatments in honey bee colonies. There are two options to limit chemical use: (1) removal of varroa from a colony by non-chemical means or (2) slowing varroa population growth [58]. Grooming and hygienic behavior within bee colonies are also important factors slowing varroa population growth [22,64]. Hygienic behavior resulted in reduced varroa mites infestation levels and therefore, this reflects in reducing varroa mites detected using brood trapping techniques [65] or sugar shake as diagnostic c methods [66]. Likewise, varroa reproduction can be further impeded by genetically selecting and maintaining mite-resistant strains of honey bees, [67–69].

One sustainable approach to mite management would be maintaining stocks of genetically resistant bees. Some bee resistance traits show potential for reducing varroa mite populations in colonies and can be used in a selective breeding program. Mite infestation may be reduced by enhancing the ability of bees to hygienically remove varroa mites from brood (i.e., Varroa sensitive hygiene (VSH)) [70,71].

Other useful heritable traits linked to reduced varroa infestation rates are active grooming of phoretic mites, post-capping duration of worker bees [72], suppression of mite reproduction, and the physical damage inflicted upon varroa by bees [68,72,73]. It is evident that proportions of non-reproducing varroa increased by 10–30% in naturally-surviving honey bee populations compared to local susceptible colonies [74]. Breeding for mite resistance is a long-term solution for varroa management in honey bee colonies. In the shorter term, beekeepers maintain strong productive colonies by rearing strong queens, performing diagnostic procedures, and using effective varroa control tactics. Therefore, treatment thresholds have been developed for specific geographic, climatic conditions adopted for specific honey bee breeds [75].

Diagnostic methods, including performing sugar shakes, water washes, and alcohol washes, are essential for evaluating mite densities in honey bee colonies. These same mite density data per adult bee or pupa when run through a life-stage model also help approximate female mite density in a given colony [76]. Assessment methods performed at different times or locations need to be comparable in order to perform effective control [19,26]. It is also evident that differences in the varroa mortality rate may be a function of bee race and climatic conditions. Reduced varroa mite longevity has been reported during winter in temperate climatic conditions [77,78]. Sampling methods including mite removal from adult bees and brood as well as counts of natural varroa mite mortality are important for establishing the level of colony infestation before implementing integrated varroa management.

There are general beekeeping practices to maintain healthy colonies by providing bees with abundant access to protein and carbohydrate. Rich pasture conditions together with regular varroa mite monitoring and organic acaricide applications can produce strong colonies more capable of overwintering and resisting disease [79]. The ultimate aim of accurate monitoring is to keep varroa mite populations well below threshold levels during the spring, summer, and autumn months by means of promoting bee hygienic behavior, drone trapping, and acaricide use [80]. As mite populations begin to rise, commercial pyrethroid-, organophosphate-, or formamidine-based acaricides can be applied [81]. It was found that one Amitraz (formamidine) fumigation was as effective as two applications of Apiguard or Thymovar [82], but efficacy decreased with time because varroa mites have become resistant to acaricides [83,84], so organic treatments are thus more useful. However, varroa populations are evolving resistance to these broad-spectrum products, whether organic or conventional [85,86]. Moreover, these products have residuals that persist in sundry hive products [6] and their formulations are not approved for organic beekeeping. As alternatives, organic substances such as essential oils and organic acids are widespread in nature, some of them occur naturally in honey, and they can be highly effective acaricides [39,55,58].

A variety of integrated varroa mite control management practices are implemented to minimize the use of broad-spectrum organophosphate- and pyrethroid-based acaricides. These practices are suitable for organic and conventional beekeeping and include such active ingredients as organic acids (e.g., formic acid, oxalic acid, and beta plant acids) and the active components of certain essential oils (e.g., thymol) [87,88]. Reliable varroa diagnoses, effective mite control, and healthy colonies are very important issues in apiculture. Therefore, integrated varroa management as a sustainable approach ensures a high quality of honey bee products with prudent use of acaricides to minimize stress on host bees. To achieve sufficient efficacy, these acaricidal substances must be applied at an optimal time to avoid collateral damage to bees and brood. Moreover, specific management practices may enhance the effectiveness of some mite control methods, including seasonal effects on efficacy, treatment protocols, active ingredient rotation, and colony sanitation [64,89]. An attempt has been made by applying 'Good Beekeeping Practice' (GBP) to reduce the use of veterinary medicines in apiculture [90]. The ultimate objective of beekeeping is to produce safe honey bee products [91]. Beekeeper activities including organized varroa control in geographical region in all beekeeping operations, beekeepers trainings and education are crucial for improving honeybee health in the context of the one health approach incorporating designing and implementing programmes, policies, legislation and research in the beekeeping sector [92].

4. Conclusions

A variety of diagnostic methods can be used to estimate varroa mite infestation in honey bee colonies, particularly in summer. Summer treatments should be conducted once the natural mite-mortality and the threshold level of the mite infestation have been determined. Differences in climatic and geographic conditions as well as hive management practices will ultimately determine the success of alternative varroa mite controls [93]. Therefore, further experiments need to be conducted in a variety of climatic conditions to determine a colony's development for specific geographical, climatic, and pasture conditions. Accurate threshold infestation levels need to be established using hive debris test, adult bee examination, or brood examination for specific geographic, climatic, and colony conditions. In addition to mite counts on hive bottom boards (mite fall), the sugar shake method (powdered sugar) helps to remove mites physically from bees and is a quick and accurate monitoring tool for assessing colony infestation levels. Coupling the powdered sugar shake method or mite falls with the use of organic acid- and botanical-based acaricides provides alternative methods of integrated varroa mite control in colonies located in both continental or warm temperate climates. Effectiveness of varroa mite control and likelihood of honey bee colonies surviving winter are further improved by using acaricides in conjunction with brood interruption or total brood removal.

Author Contributions: A.G., and B.S. equally contributed to all studies and wrote the manuscript.

Funding: This research was financially supported by the Slovenian Research Agency, Research Program P1-0164; Research for improvement of safe food and health.

Acknowledgments: Research topic is part of Varroa control Task Force activities. Coloss is Association, which aims to explain and prevent massive honey bee colony losses. (Prevention of honey bee COlony LOSSes).

Conflicts of Interest: The authors declare no conflicts of interest.

References

1. Jong, D.D.; Jong, P.H.D.; Gonçalves, L.S. Weight Loss and Other Damage to Developing Worker Honeybees from Infestation with Varroa Jacobsoni. *J. Apic. Res.* **1982**, *21*, 165–167. [CrossRef]
2. Martin, S.J. Ontogenesis of the mite Varroa jacobsoni Oud. in worker brood of the honeybee Apis mellifera L. under natural conditions. *Exp. Appl. Acarol.* **1994**, *18*, 87–100. [CrossRef]
3. Büchler, R. Varroa Tolerance in Honey Bees—Occurrence, Characters and Breeding. *Bee World* **2015**, *75*, 54–70. [CrossRef]
4. Roberts, J.M.K.; Anderson, D.L.; Durr, P.A. Absence of deformed wing virus and Varroa destructor in Australia provides unique perspectives on honeybee viral landscapes and colony losses. *Sci. Rep.* **2017**, *7*, 1–11. [CrossRef]
5. McMenamin, A.J.; Genersch, E. Honey bee colony losses and associated viruses. *Curr. Opin. Insect Sci.* **2015**, *8*, 121–129. [CrossRef]
6. Wallner, K. Varroacides and their residues in bee products. *Apidologie* **1999**, *30*, 235–248. [CrossRef]
7. Ruijter, A. Issues in the control of Varroa infestation. In *New Perspectives on Varroa*; Mathenson, A., Ed.; IBRA: Cardiff, UK, 1994; pp. 24–26.
8. Lee, K.V.; Moon, R.D.; Burkness, E.C.; Hutchison, W.D.; Spivak, M. Practical sampling plans for Varroa destructor (Acari: Varroidae) in Apis mellifera (Hymenoptera: Apidae) colonies and apiaries. *J. Econ. Entomol.* **2010**, *103*, 1039–1050. [CrossRef]
9. OIE (World Organisation for Animal Health). Varoosis of honey bees. Chapter 2.2.7. In *OIE. Manual of Diagnostic Tests and Vaccines for Terrestrial Animals (Mammals, Birds and Bees)*; OIE: Paris, France, 2008; Volume 6, pp. 424–430.
10. Ritter, W. Varroa Disease of the Honeybee Apis Mellifera. *Bee World* **1981**, *62*, 141–153. [CrossRef]
11. Gregorc, A.; Jelenc, J. Control of Varroa Jacobsoni Oud. in honeybee colonies using Apilife-Var. *Zb. Vet. Fak. Univ. Ljubljana* **1996**, *33*, 231–235.
12. Liebig, G.; Schlipf, U.; Fremuth, W.; Ludwig, W. Ergebnisse der untersuchungen über die befallsentwicklung der Varroa-Milbe in Stuttgart-Hohenheim 1983. *Allg. Dtsch. Imkerztg.* **1984**, *18*, 185–190.

13. Harris, J. Managing Varroa Mites in Honey Bee Colonies | Mississippi State University Extension Service. Available online: http://extension.msstate.edu/publications/managing-varroa-mites-honey-bee-colonies (accessed on 9 December 2019).

14. Mattila, H.R.; Otis, G.W. The efficacy of Apiguard against varroa and tracheal mites, and its effect on honey production: 1999 trial. *Am. Bee J.* **2000**, *140*, 969–973.

15. Martin, S. A population model for the ectoparasitic mite Varroa jacobsoni in honey bee (Apis mellifera) colonies. Ecological Modelling 109. *Ecol. Model.* **1998**, *109*, 267–281. [CrossRef]

16. Branco, M.R.; Kidd, N.A.; Pickard, R.S. A comparative evaluation of sampling methods for Varroa destructor (Acari: Varroidae) population estimation. *Apidologie* **2006**, *37*, 452–461. [CrossRef]

17. Ritter, W. Neuester Stand der diagnostischen und terapeutischen Massnahmen zur Bekämpfung der Varroatose. *Tierarztl. Umsch.* **1984**, *39*, 122–127.

18. Spivak, M.; Reuter, G.S. Varroa destructor Infestation in Untreated Honey Bee (Hymenoptera: Apidae) Colonies Selected for Hygienic Behavior. *J. Econ. Entomol.* **2001**, *94*, 326–331. [CrossRef]

19. Fakhimzadeh, K.; Ellis, J.D.; Hayes, J.W. Physical control of varroa mites (Varroa destructor): the effects of various dust materials on varroa mite fall from adult honey bees (Apis mellifera) in vitro. *J. Apic. Res.* **2011**, *50*, 203–211. [CrossRef]

20. Fakhimzadeh, K. The effects of powdered sugar varroa control treatments on Apis mellifera colony development. *J. Apic. Res.* **2001**, *40*, 105–109. [CrossRef]

21. Aliano, N.P.; Ellis, M.D. A strategy for using powdered sugar to reduce varroa populations in honey bee colonies. *J. Apic. Res.* **2005**, *44*, 54–57. [CrossRef]

22. Macedo, P.A.; Wu, J.; Ellis, M.D. Using inert dusts to detect and assess varroa infestations in honey bee colonies. *J. Apic. Res.* **2002**, *41*, 3–7. [CrossRef]

23. Dietemann, V.; Nazzi, F.; Martin, S.J.; Anderson, D.L.; Locke, B.; Delaplane, K.S.; Wauquiez, Q.; Tannahill, C.; Frey, E.; Ziegelmann, B.; et al. Standard methods for varroa research. *J. Apic. Res.* **2013**, *52*, 1–54. [CrossRef]

24. Gregorc, A.; Alburaki, M.; Sampson, B.; Knight, P.R.; Adamczyk, J. Toxicity of Selected Acaricides to Honey Bees (Apis mellifera) and Varroa (Varroa destructor Anderson and Trueman) and Their Use in Controlling Varroa within Honey Bee Colonies. *Insects* **2018**, *9*, 55. [CrossRef] [PubMed]

25. Gregorc, A.; Alburaki, M.; Werle, C.; Knight, P.R.; Adamczyk, J. Brood removal or queen caging combined with oxalic acid treatment to control varroa mites (Varroa destructor) in honey bee colonies (Apis mellifera). *Apidologie* **2017**, *48*, 821–832. [CrossRef]

26. Toufailia, H.M.A.; Amiri, E.; Scandian, L.; Kryger, P.; Ratnieks, F.L.W. Towards integrated control of varroa: effect of variation in hygienic behaviour among honey bee colonies on mite population increase and deformed wing virus incidence. *J. Apic. Res.* **2014**, *53*, 555–562. [CrossRef]

27. Medina, L.M.; Martin, S.J. A Comparative Study of Varroa Jacobsoni Reproduction in Worker Cells of Honey Bees (Apis Mellifera) in England and Africanized Bees in Yucatan, Mexico. *Exp. Appl. Acarol.* **1999**, *23*, 659–667. [CrossRef]

28. Lindberg, C.M.; Melathopoulos, A.P.; Winston, M.L. Laboratory evaluation of miticides to control Varroa jacobsoni (Acari: Varroidae), a honey bee (Hymenoptera: Apidae) parasite. *J. Econ. Entomol.* **2000**, *93*, 189–198. [CrossRef] [PubMed]

29. Imdorf, A.; Bogdanov, S.; Ibáñez Ochoa, R.; Calderone, N.W. Use of essential oils for the control of Varroa jacobsoni Oud. in honey bee colonies. *Apidologie* **1999**, *30*, 209–228. [CrossRef]

30. Giacomelli, A.; Pietropaoli, M.; Carvelli, A.; Iacoponi, F.; Formato, G. Combination of thymol treatment (Apiguard®) and caging the queen technique to fight Varroa destructor. *Apidologie* **2016**, *47*, 606–616. [CrossRef]

31. Bogdanov, S.; Kilchenmann, V.; Imdorf, A.; Fluri, P. Residues in honey after application of thymol against varroa using the Frakno Thymol Frame (Reprinted from Schweizerisch Bienen-Zeitung, vol 121, pg 224–226, 1998). *Am. Bee J.* **1998**, *138*, 610–611.

32. Nanetti, A.; Stradi, G. Varroasi: trattamento chimico con acido ossalico in sciroppo zuccherino. *L'APE Nostra Amica* **1997**, *5*, 6–14.

33. Fassbinder-Orth, C.; Grodnitzky, J.; Coats, J. Monoterpenoids as possible control agents for Varroa destructor. *J. Apic. Res.* **2002**, *41*, 83–88. [CrossRef]

34. Melathopoulos, A.P.; Gates, J. Comparison of two thymol-based acaricides, Api Life Var (R) and Apiguard (TM), for the control of Varroa mites. *Am. Bee J.* **2003**, *43*, 489–493.

35. Coffey, M.F.; Breen, J. Efficacy of Apilife Var® and Thymovar® against Varroa destructor as an autumn treatment in a cool climate. *J. Apic. Res.* **2013**, *52*, 210–218. [CrossRef]

36. Gregorc, A.; Adamczyk, J.; Kapun, S.; Planinc, I. Integrated varroa control in honey bee (Apis mellifera carnica) colonies with or without brood. *J. Apic. Res.* **2016**, *55*, 253–258. [CrossRef]

37. DeGrandi-Hoffman, G.; Ahumada, F.; Probasco, G.; Schantz, L. The effects of beta acids from hops (Humulus lupulus) on mortality of Varroa destructor (Acari: Varroidae). *Exp. Appl. Acarol.* **2012**, *58*, 407–421. [CrossRef] [PubMed]

38. Rademacher, E.; Harz, M.; Schneider, S. The development of HopGuard® as a winter treatment against Varroa destructor in colonies of Apis mellifera. *Apidologie* **2015**, *46*, 748–759. [CrossRef]

39. Wilson, W.; Collins, A. Failure of formic acid to control Varroa jacobsoni in a hot climate. *Am. Bee J.* **1993**, *133*, 863–872.

40. Fries, I. Treatment of sealed honey bee brood with formic acid for control of Varroa jacobsoni. *Am. Bee J. (USA)* **1991**, *131*, 313–314.

41. Hoppe, H.; Ritter, W.; Stephen, E.W.C. The control of parasitic bee mites: Varroa jacobsoni, Acarapis woodi, and Tropilaelaps clarae with formic acid. *Am. Bee J.* **1989**, *129*, 739–742.

42. Imdorf, A.; Charrière, J.-D.; Maquelin, C.; Kilchenmann, V.; Bachofen, B. Alternative Varroa control. *Am. Bee J.* **1996**, *136*, 189–193.

43. Liu, T.P. Formic acid and bee mites. *Am. Bee J. (USA)* **1991**, *131*, 311–312.

44. Nanetti, A. *Oxalic Acid for Mite Control—Results and Review*; Coordination in Europe of research on integrated control of Varroa mites in honey bee colonies; Commission of the European Communities: Brussels, Belgium; Luxembourg, 1999; pp. 7–14.

45. Charriére, J.-D.; Imdorf, A. Oxalic acid treatment by trickling against Varroa destructor: recommendations for use in central Europe and under temperate climate conditions. *Bee World* **2002**, *83*, 51–60. [CrossRef]

46. Gregorc, A.; Planinc, I. Dynamics of Falling Varroa Mites in Honeybee (Apis mellifera) Colonies Following Oxalic Acid Treatments. *Acta Vet. Brno* **2004**, *73*, 385–391.

47. Aliano, N.P.; Ellis, M.D.; Siegfried, B.D. Acute contact toxicity of oxalic acid to Varroa destructor (Acari: Varroidae) and their Apis mellifera (Hymenoptera: Apidae) hosts in laboratory bioassays. *J. Econ. Entomol.* **2006**, *99*, 1579–1582. [CrossRef] [PubMed]

48. Brødsgaard, C.J.; Jansen, S.E.; Hansen, C.W.; Hansen, H. Spring treatment with oxalic acid in honey bee colonies as varroa control. *DIAS Rep. Horticult.* **1999**, *6*, 16.

49. Gregorc, A.; Planinc, I. Acaricidal effect of oxalic acid in honeybee (apis mellifera) colonies. *Apidologie* **2001**, *32*, 333–340. [CrossRef]

50. Nanetti, A.; Massi, S.; Mutinelli, F.; Cremasco, S. L'acido ossalico nel controllo della varroasi: Note preliminari. *Apitalia* **1995**, *22*, 29–32.

51. Imdorf, A.; Charrière, J.-D.; Bachofen, B. Efficiency checking of the Varroa jacobsoni control methods by means of oxalic acid. *Apiacta* **1997**, *32*, 89–91.

52. Mutinelli, F.; Baggio, A.; Capolongo, F.; Piro, R.; Prandin, L.; Biasion, L. A scientific note on oxalic acid by topical application for the control of varroosis. *Apidologie* **1997**, *28*, 461–462. [CrossRef]

53. Toufailia, H.A.; Ratnieks, F.L.W. How effective is Apistan® at killing varroa? Results from a LASI trial. *Bee Craft* **2016**, *98*, 7–11.

54. Bacandritsos, N.; Papanastasiou, I.; Saitanis, C.; Nanetti, A.; Roinioti, E. Efficacy of repeated trickle applications of oxalic acid in syrup for varroosis control in Apis mellifera: influence of meteorological conditions and presence of brood. *Vet. Parasitol.* **2007**, *148*, 174–178. [CrossRef]

55. Toufailia, H.A.; Scandian, L.; Ratnieks, F.L.W. Towards integrated control of varroa: 2)comparing application methods and doses of oxalic acid on the mortality of phoretic Varroa destructor mites and their honey bee hosts. *J. Apic. Res.* **2015**, *54*, 108–120. [CrossRef]

56. Floris, I.; Satta, A.; Mutinelli, F.; Pradin, L. Efficacy of winter applications of oxalic acid against Varroa jacobsoni Oudemans in beehives in a Mediterranean area (Sardinia, Italy). *Redia* **1998**, *81*, 143–150.

57. Vetharaniam, I. Predicting reproduction rate of varroa. *Ecol. Model.* **2012**, *224*, 11–17. [CrossRef]

58. Delaplane, K.S.; Berry, J.A.; Skinner, J.A.; Parkman, J.P.; Hood, W.M. Integrated pest management against Varroa destructor reduces colony mite levels and delays treatment threshold. *J. Apic. Res.* **2005**, *44*, 157–162. [CrossRef]

59. Rosenkranz, P.; Aumeier, P.; Ziegelmann, B. Biology and control of Varroa destructor. *J. Invertebr. Pathol.* **2010**, *103*, 96–119. [CrossRef]

60. Nanetti, A.; Higes, M.; Baracani, G.; Besana, A. Control de la varroa con ácido oxálico en combinación con la interrupción artificial de la puesta de cría. In Proceedings of the II Congreso Ibérico de Apicultura, Guadalajara, Spain, 24 October 2012; pp. 18–20.

61. Pietropaoli, M.; Giacomelli, A.; Milito, M.; Pizzariello, M.; Carla, G.; Scholl, F.; Formato, G. Integrated Pest Management strategies against Varroa destructor: the use of oxalic acid combined with innovative cages to obtain the absence of brood. *Eur. J. Integr. Med.* **2013**, *15*, 93.

62. Calis, J.N.M.; Boot, W.J.; Beetsma, J.; van den Eijnde, J.H.P.M.; de Ruijter, A.; van der Steen, J.J.M. Effective biotechnical control of varroa: applying knowledge on brood cell invasion to trap honey bee parasites in drone brood. *J. Apic. Res.* **1999**, *38*, 49–61. [CrossRef]

63. Maggi, M.; Tourn, E.; Negri, P.; Szawarski, N.; Marconi, A.; Gallez, L.; Medici, S.; Ruffinengo, S.; Brasesco, C.; De Feudis, L.; et al. A new formulation of oxalic acid for Varroa destructor control applied in Apis mellifera colonies in the presence of brood. *Apidologie* **2016**, *47*, 596–605. [CrossRef]

64. Spivak, M.; Gilliam, M. Hygienic behaviour of honey bees and its application for control of brood diseases and varroa: Part II. Studies on hygienic behaviour since the Rothenbuhler era. *BEE World* **1998**, *79*, 169–186. [CrossRef]

65. Schulz, A.E. Reproduktion und populationsentwicklung der parasitischen milbe varroa jacobsoni oud. In abhängigkeit vom brutzyklus ihres wirtes apis mellifera l. (i. Teil). *Apidologie* **1984**, *15*, 401–420. [CrossRef]

66. Fakhimzadeh, K. Potential of Super-Fine Ground, Plain White Sugar Dusting as An Ecological Tool for the Control of Varroasis in the Honey Bee (Apis mellifera). Available online: https://eurekamag.com/research/003/530/003530144.php (accessed on 19 October 2019).

67. Spivak, M. Honey bee hygienic behavior and defense against Varroa jacobsoni. *Apidologie* **1996**, *27*, 245–260. [CrossRef]

68. Harbo, J.R.; Harris, J.W. Selecting honey bees for resistance to Varroa jacobsoni. *Apidologie* **1999**, *30*, 183–196. [CrossRef]

69. Harbo, J.R.; Hoopingarner, R.A. Honey bees (Hymenoptera: Apidae) in the United States that express resistance to Varroa jacobsoni (Mesostigmata: Varroidae). *J. Econ. Entomol. (USA)* **1997**, *90*, 893–898. [CrossRef]

70. Danka, R.G.; Harris, J.W.; Villa, J.D. Expression of Varroa Sensitive Hygiene (VSH) in Commercial VSH Honey Bees (Hymenoptera: Apidae). *J. Econ. Entomol.* **2011**, *104*, 745–749. [CrossRef]

71. Ibrahim, A.; Spivak, M. The relationship between hygienic behavior and suppression of mite reproduction as honey bee (Apis mellifera) mechanisms of resistance to Varroa destructor. *Apidologie* **2006**, *37*, 31–40. [CrossRef]

72. Moritz, R.F.A. Heritability of the postcapping stage in Apis mellifera and its relation to varroatosis resistance. *J. Hered.* **1985**, *76*, 267–270. [CrossRef]

73. Fries, I.; Camazine, S.; Sneyd, J. Population Dynamics of Varroa Jacobsoni: A Model and a Review. *Bee World* **1994**, *75*, 5–28. [CrossRef]

74. Oddie, M.; Büchler, R.; Dahle, B.; Kovacic, M.; Le Conte, Y.; Locke, B.; de Miranda, J.R.; Mondet, F.; Neumann, P. Rapid parallel evolution overcomes global honey bee parasite. *Sci. Rep.* **2018**, *8*, 7704. [CrossRef]

75. Locke, B. Natural Varroa mite-surviving Apis mellifera honeybee populations. *Apidologie* **2016**, *47*, 467–482. [CrossRef]

76. Martin, S. Population modelling and the production of a monitoring tool for Varroa jacobsoni an ectoparasitic mite of honey bees. *Asp. Appl. Biol.* **1999**, *53*, 105–112.

77. Moritz, R.F.A.; Mautz, D. Development of Varroa jacobsoni in colonies of Apis mellifera capensis and Apis mellifera carnica. *Apidologie* **1990**, *21*, 53–58. [CrossRef]

78. Ruttner, F.; Hänel, H. Active defense against Varroa mites in a Carniolan strain of honeybee (Apis mellifera carnica Pollmann). *Apidologie* **1992**, *23*, 173–187. [CrossRef]

79. Giovenazzo, P.; Dubreuil, P. Evaluation of spring organic treatments against Varroa destructor (Acari: Varroidae) in honey bee Apis mellifera (Hymenoptera: Apidae) colonies in eastern Canada. *Exp. Appl. Acarol.* **2011**, *55*, 65–76. [CrossRef] [PubMed]

80. Genersch, E. Honey bee pathology: Current threats to honey bees and beekeeping. *Appl. Microbiol. Biotechnol.* **2010**, *87*, 87–97. [CrossRef] [PubMed]

81. Norain Sajid, Z.; Aziz, M.A.; Bodlah, I.; Rana, R.M.; Ghramh, H.A.; Khan, K.A. Efficacy assessment of soft and hard acaricides against Varroa destructor mite infesting honey bee (Apis mellifera) colonies, through sugar roll method. *Saudi J. Biol. Sci.* **2019**. [CrossRef]

82. Gregorc, A.; Planinc, I. Use of Thymol Formulations, Amitraz, and Oxalic Acid for the Control of the Varroa Mite in Honey Bee (Apis mellifera carnica) Colonies. *J. Apicult. Sci.* **2012**, *56*, 61–69. [CrossRef]

83. Sammataro, D.; Untalan, P.; Guerrero, F.; Finley, J. The resistance of varroa mites (Acari: Varroidae) to acaricides and the presence of esterase. *Int. J. Acarol.* **2005**, *31*, 67–74. [CrossRef]

84. Mathieu, L.; Faucon, J.-P. Changes in the response time for Varroa jacobsoni exposed to amitraz. *J. Apic. Res.* **2000**, *39*, 155–158. [CrossRef]

85. Mozes-Koch, R.; Slabezki, Y.; Efrat, H.; Kalev, H.; Kamer, Y.; Yakobson, B.A.; Dag, A. First detection in Israel of fluvalinate resistance in the varroa mite using bioassay and biochemical methods. *Exp. Appl. Acarol.* **2000**, *24*, 35–43. [CrossRef]

86. Floris, I.; Cabras, P.; Garau, V.L.; Minelli, E.V.; Satta, A.; Troullier, J. Persistence and Effectiveness of Pyrethroids in Plastic Strips Against Varroa jacobsoni (Acari: Varroidae) and Mite Resistance in a Mediterranean Area. *J. Econ. Entomol.* **2001**, *94*, 806–810. [CrossRef]

87. Imdorf, A.; Charrière, J.-D.; Kilchenmann, V.; Bogdanov, S.; Fluri, P. Alternative strategy in central Europe for the control of Varroa destructor in honey bee colonies. *Apiacta* **2003**, *38*, 258–285.

88. Calderone, N. Evaluation of Drone Brood Removal for Management of Varroa destructor (Acari: Varroidae) in Colonies of Apis mellifera (Hymenoptera: Apidae) in the Northeastern United States. *J. Econ. Entomol.* **2005**, *98*, 645–650. [CrossRef] [PubMed]

89. Currie, R.W.; Gatien, P. Timing acaricide treatments to prevent Varroa destructor (Acari: Varroidae) from causing economic damage to honey bee colonies. *Can. Entomol.* **2006**, *138*, 238–252. [CrossRef]

90. Beekeeping | TECA—Technologies and Practices for Small Agricultural Producers | Food and Agriculture Organization of the United Nations. Available online: http://www.fao.org/teca/forum/beekeeping/en (accessed on 22 October 2019).

91. D'Ascenzi, C.; Formato, G.; Martin, P. Chemical hazards in honey. In *Chemical Hazards in Foods of Animal Origin*; ECVPH Food safety assurance; Wageningen Academic Publishers: Wageningen, The Netherlands, 2018; Volume 7, pp. 443–475. ISBN 978-90-8686-326-6.

92. WHO | One Health. Available online: http://www.who.int/features/qa/one-health/en/ (accessed on 22 October 2019).

93. Trouiller, J.; Watkins, M. Experimentation on Apiguard—A controlled release gel of thymol against honey bee diseases. In Proceedings of the 37th International Apicultural Congress, Durban, South Africa, 28 October–1 November 2001.

Review

Biotic and Abiotic Factors Associated with Colonies Mortalities of Managed Honey Bee (*Apis mellifera*)

Boyko Neov [1], Ani Georgieva [2], Rositsa Shumkova [3], Georgi Radoslavov [1] and Peter Hristov [1,*]

[1] Department of Animal Diversity and Resources, Institute of Biodiversity and Ecosystem Research, Bulgarian Academy of Sciences, "Acad. G. Bonchev" Str., Bl. 25, Sofia 1113, Bulgaria; boikoneov@gmail.com (B.N.); gradoslavov@gmail.com (G.R.)

[2] Department of Pathology, Institute of Experimental Morphology, Pathology and Morphology and Anthropology with Museum, Bulgarian Academy of Sciences, Sofia 1113, Bulgaria; georgieva_any@abv.bg

[3] Research Centre of Stockbreeding and Agriculture, Agricultural Academy, Smolyan 4700, Bulgaria; rositsa6z@abv.bg

* Correspondence: peter_hristoff@abv.bg; Tel.: +35-929-792-327

Received: 2 November 2019; Accepted: 9 December 2019; Published: 10 December 2019

Abstract: Despite the presence of a large number of pollinators of flowering plants worldwide, the European honey bee, *Apis melifera*, plays the most important role in the pollination of a number of crops, including all vegetables, non-food crops and oilseed crops, decorative and medical plants, and others. The experience of isolated cases of complete extinction of honey bees in individual regions has shown that this phenomenon leads to a dramatic pollination crisis and reduced ability or even total inability to grow insect-pollinated crops if relying solely on native, naturally occurring pollinators. Current scientific data indicate that the global bee extinction between the Cretaceous and the Paleogene (Cretaceous-Tertiary) occurred, which led to the disappearance of flowers because they could not produce viable fruit and germinate due to lack of pollination by bees or other animals. From the Middle Ages to the present day, there has been evidence that honey bees have always overcome the adverse factors affecting them throughout the ages, after which their population has fully recovered. This fact must be treated with great care given the emergence of a new, widespread stress factor in the second half of the 20th century—intoxication of beehives with antibiotics and acaricides, and treatment of crops with pesticides. Along with acute and chronic intoxication of bees and bee products, there are other new major stressors of global importance reducing the number of bee colonies: widespread prevalence of pathogenic organisms and pest beetles, climate change and adverse climatic conditions, landscape changes and limitation of natural habitats, intensification of agricultural production, inadequate nutrition, and introduction of invasive species. This report summarizes the impact of individual negative factors on the health and behavior of bees to limit the combined effects of the above stressors.

Keywords: *Apis mellifera*; honey bee colony losses; biotic factors; abiotic factors

---

1. Introduction

The European honey bee, *Apis mellifera* L. (Hymenoptera: Apidae) has been managed for millennia [1]. The species was domesticated by the ancient Egyptians around 2600 BC [2]. Since then, bees have been used as a source of unique, natural, multifunctional products (basically honey and wax). However, the more significant ecological function of the species is related to the pollination of a wide range of agricultural and wild plant species. The honey bee is one of about 200,000 existing species of plant pollinators [3]. Approximately 84% of the crops grown in Europe are dependent on insect pollination [4]. The only major exception is grain crops that are wind-pollinated. Bees, both wild and managed, provide about 80% of the insect pollination of flowering plants [5]. Among them, the honey bee plays the most important role in the process, especially in the pollination of apple, cherry, small grapes, kiwi, broccoli, carrots, cauliflower, celery, cucumber, onion, pumpkin, legumes, sunflower, and many others.

The role of *A. mellifera* as a pollinator in natural habitats is important for several reasons. First, animal-mediated pollination represents vital ecosystem maintenance, i.e., 87.5% of flowering plant species are pollinated by animals [6]. It is well known that the honey bee is considered as a super generalist, which provides a vital role in the functioning of many terrestrial ecosystems [7]. Unlike other pollinators, which have declined significantly in recent years, mainly as a result of habitat loss, *A. mellifera* populations can overcome these disorders [8]. The recent increases in the mortality of managed *A. mellifera* colonies in some parts of the world (USA) may increase the feral population, which is more susceptible mainly to Varroa mite [9]. In some countries, the introduction of populations *A. mellifera* has led to competition with other pollinators [10,11]. A typical example is Australia, where beekeepers are restricted to use agricultural land but not national parks for this very reason [12]. Given the importance of honey bees for the nourishment of humankind, the reducing of the number of bee colonies in Europe and America has been met with increasing concern in recent decades. Some of the published materials describe the unusual and somewhat tragic consequences of a future crisis in honey bee populations and other pollinator species [13,14]. Other studies have linked the problem with the degradation of natural ecosystems and the persistent tendency to lose ever-increasing amounts of the planet's biodiversity [15,16].

The purpose of this report is to summarize and analyze data published in the scientific literature on the causes, drivers, and conditions associated with the poor health status and death of bees and bee colonies.

2. Historical Occurrences of Honey Bee Decline

2.1. Prehistoric Extinction Event

There have been known cases of complete bee extinction or population crises. It has been established that prehistoric mass extinction of bee species from the Xylocopinae colony took place during the transition between the Cretaceous and Paleogene geological periods, about 65.5 million years ago [17]. The period has left a characteristic imprint on geological layers—a thin clay layer known as the K–Pg (Cretaceous-Paleogene) or K–T boundary. On the base of sequence and phylogenetic analysis of two mitochondrial COI (cytochrome oxidase subunit 1) and cytb (cytochrome b) and two nuclear genes, F1 and F2 Elongation Factor-1α (EF-1α) of Xylocopinae tribe [17], the authors suggested the disruption of many plant-insect relationships at that time, which had a negative effect on the bees. The Cretaceous–Paleogene mass extinction event is believed to have been caused by a devastating change in the global climate.

2.2. Medieval Bee Mortalities

In medieval Europe, there was a mention of "bee mortality" in Ireland in 951, according to the Annals of Ulster. Citing the same source, Fleming [18] noted "high mortality among humans, cattle and bees" again in Ireland in 992. In the same year, the extinction of bees was preceded by a long and severe winter, and drought in the summer, combined with an attack by fungal "diseases or turf" (namely, ergot, *Claviceps purpurea* (Fr.) Tul. 1853) on cereals. The ergot sclerotium contains high concentrations (up to 2% of dry mass) of the alkaloid ergotamine, which accumulates in the so-called sugary honeydew, which attracts bees. The result was a mass famine and an epidemic of ergotism in France in the same year. The harsh winter and dry summer preceded the "unprecedented pestilence and devastation of the bees" affecting all of Bavaria in 1035. The unusual cold in the summer of that year eliminated the production of corn and fruit (Annals of Ulster).

Even these historical records from medieval Europe point to a clear causality: bad weather resulting in decreased nectar production, which, in turn, had a negative impact on colony productivity.

The earliest recorded shipment of bees to the New World (North America) took place from England in 1621. By 1650, nearly all farms in New England are reported to have had one or two colonies of bees, but there is evidence that the number of bees managed by American colonists declined after 1670, 50 years after the honey bee was introduced from Europe to North America. According to Pellett [19], the most likely cause was associated with an unknown infectious disease.

2.3. Other Unexplained Evidence for Bee Mortality

Many of the bee epidemics recorded in detail in the past remain unexplained. For example, in 1903, 2000 colonies were lost to an unknown "disappearing disease" in the Cache Valley of Utah [20]. Significantly, however, the previous winter had been hard and the spring cold. In 1995–1996, beekeepers in Pennsylvania lost 53% of their colonies to no specifically identifiable cause [21]. One of the most famous epidemics occurred on the Isle of Wight, a small island off the south coast of England, when three separate epidemics between 1905 and 1919 wiped out 90% of the island's bees [22]. In this case, pathogenic factors, such as the tracheal mite *Acarapis woodi* and the microsporidia *Nosema apis*, were involved. Thereafter, after long and lengthy research over the years it is suggested that the disease had been due to a combination of factors, in particular, infection by Chronic Bee Paralysis Virus (completely unknown at the time), together with unfavorable climatic conditions that restricted the growth of plants and, hence, indirectly affected the foraging by bees because of the lack of flowers [23].

Although the number of managed beehives globally has increased by 45% since 1961, the proportion of bee-dependent pollinating crops has increased much more steeply—by 300% [24]. In this regard, many beehives, especially in countries with industrialized, high-intensity agriculture, are raised as mobile units for the purpose of pollination of crops, not for the production of bee honey and other products [25]. Insufficient attention has been paid to the potential negative effect of long-distance transport of beehives and the related stress for bees [26,27].

3. Colony Collapse Disorder Syndrome

Many studies link the death of bee colonies to the popular Colony Collapse Disorder symptom/syndrome (CCD). As the name suggests, a typical characteristic of this pathological phenomenon, is the total absence or the presence of very few adult (imagined) bees in the hive, in the presence of live brood (eggs, larvae, pre-pupae, and pupae) and an alive bee queen. A common feature is that worker bees leave the hive and do not return [28,29]. In fact, there is no factual evidence of the death of individuals. No dead bodies are found inside or in front of the affected hive, which is a characteristic sign of acute poisoning [30]. In addition, there is no indication of an invasion of the hive by other insects, such as wasps or hornets. Visual inspection shows that live individuals in the hive, both imagined and brood, are usually infested by the ectoparasitic mite—*Varroa destructor* [31]. Other pests found in all bee colonies affected by the syndrome are intracellular, intestinal parasites (microsporidia), mainly *Nosema ceranae*, and other representatives of the same genus [32]. Other unspecific signs indicating that the bee colonies are in crisis but not necessarily dictated by CCD include the shortened period (from approximately 25 to 5 days) during which young bees take care of larvae and pupae before becoming workers that produce wax; significantly delayed attacks by hive pests, colony includes mostly young adult bees, etc. In its classic form, the syndrome appears to be localized only in the United States and, in some cases, in Europe [33].

4. What Actually Leads to Honey Bee Decline?

Widespread notions, including such among the academic community, relate the main problem—the decrease in the number of bee colonies and any other form of bee losses—mainly or exclusively, to the aforementioned pathological phenomenon of CCD. However, scientists and experts involved in researching the problem consider two forms of honey bee losses:

1. Annual (most frequent)—as a result of unsuccessful wintering caused by biotic factors (such as infections and parasites), acute intoxication, and a number of other causes, which are subject of the discussion in this review [34]

2. Multi-annual—permanent reduction in the number of bee colonies in separate, specific regions.

For example, it is known that from the mid-1980s until now, the number of bee colonies in Europe has decreased by 25% and in the United States by 50–60% [35]. At the same time, despite sporadic local extinction, the number of bee colonies globally has increased by about 45% over the last 50 years and more specifically since 1961 [24,35]. The latter seems hopeful, but these data must be interpreted with care. Usually, beekeepers compensate for the loss of one hive by splitting it into two. This results in the "weakening" of the two new bee colonies, and the long-term effect of repeated application of this practice has not been fully studied. Very often, there is a lack of objectively collected information on the status of the honey bee in individual countries, on population changes over the years, and by region. For example, in Austria and Czechia, colony losses during winter were fluctuating from year to year, with strong regional differences. It has been observed that winter losses related to queen problems, differences in population dynamics, and treatment against the Varroa mite, differences in the number of beekeepers and colonies, etc. [36]. Generally, in the 20th century in Europe, two periods were distinguished according to the prevailing trend: an increase in the total number of bee colonies in 1965–1985, and a decrease in the number of beehives in 1986–2005 [4]. The number of beekeepers, both professionally engaged and those for whom beekeeping is a hobby (holdings with up to 50 bee colonies), has also been decreasing in Europe. The changes recorded between 2000 and 2010 show an increase in the total number of hives in Europe from 15 million in 2000 to 16.4 million in 2007, a decrease in 2008–2009, and again, some increase towards the end of the period—2010, to about 15.8 million hives [37]. In comparison, in 2010, the registered bee colonies in Bulgaria were over 613,000 [35]. While, in 2006, bee colonies in the country amounted to 671,674, in 2007 to 718,822, in the following years, their number kept steadily decreasing, and in 2012 it reached 529,117—a reduction from 21% to 26.4%. During this period, the number of holdings engaged in beekeeping (beekeepers) also decreased from 31,026 in 2008 to 19,238 in 2012—a 40% reduction.

5. Factors Affecting the Health of Honey Bees

There are five stressors of global importance that are thought to be relevant to the reducing number of bee colonies in different parts of the world. These are the anthropogenic driven worldwide spread of pathogenic organisms and pest beetles (*Aethina tumida*), climate change and adverse climatic conditions, landscape changes with limitation of natural habitats, intensification of agricultural production (including the use of fertilizers and pesticides), and invasion of new non-native species [38].

5.1. Biotic Factors

A total of 29 diseases and pests are known to be the cause of the annual loss of bee colonies almost everywhere in the world [35]. Among the causative agents and pests in the honey bee colonies are Varro mite (*Varroa destructor*), microsporidia (*Nosem apis*; *N. ceranae*, the more virulent one); fungi such as *Ascophaera apis*; bacteria (*Paenibacillus larvae*, *Melissococcus plutonius*), amoebae (*Malpighamoeba mellificae*), septicemia and spiroplasm, small hive beetles (*Aethina tumida*), wax moths (Pyralidae), and others [3,38].

The parasitic mite *Varroa destructor* was introduced into Europe (possible in the 1950s) [39], and North America (first detected in 1987s), [40], making it a nearly ubiquitous honey bee pest [41]. An exception is Australia, as well as some isolated islands and probably individual Central African countries where the parasite has not yet been identified [4]. This mite is responsible for significant annual losses of bee colonies in Canada and a number of European countries [35]. In regions where *V. destructor* is not a major problem, beekeeping is negligible or does not occur at all [42]. A number of authors have identified this species as the most probable cause of bee populations extinction in Europe, USA, and Canada [31,35,43–45]. In addition to being direct pests, mites are a vector of various viral diseases in parasitized bees [28,42]. Examples of the latter include the Deformed Wings Virus, the Acute Bee Paralysis Virus, the Israeli Acute Paralysis Virus, and the Kashmir Bee Virus—all directly related to the empty hive syndrome [46–48].

Microsporidian infections with Nosema spp.—*N. ceranae* and *N. apis*—lead to acute diarrhea in bees during the winter-spring season or to a latent infection [49,50]. There have been cases where infections with *N. ceranae* cause symptoms identical to the decline of bee colonies [51]), but the relationship between the two pathological phenomena are poorly understood [35].

The small hive beetle, *Aethina tumida*, adversely affects all aspects of beekeeping, including queen rearing, honey production and processing, and pollination operations. In honey bee colonies, they feed on pollen, honey, and occasionally brood. The damage associated with an *Aethina tumida* infestation is caused by the beetle larval stage; adults have little negative impact on a colony besides distracting worker bees from their normal hive duties [52].

The Greater wax moth (*Galleria mellonella*) is another opportunistic pest found in honey beehives, and cause significant damage to stored combs [53]. The damage caused by *G. mellonella* larvae is severe in tropical and sub-tropical regions and is believed to be one of the contributing factors to the decline in both feral and wild honeybee populations.

While in Europe, biotic factors, mainly infections and parasites of the hives, are given great importance; most of the loss of bee colonies in the United States is attributed to adverse weather factors, starvation, loss of the bee queen or stress from transporting hives over long distances, and infection (mainly viruses, bacteria, and fungi) that may contribute to CCD syndrome [27,29,54,55]. However, the studies cited here do not take into account the prior health status of beehives killed in case of transport stress [35].

5.2. Abiotic Factors

5.2.1. Climate Change

In order to assess the potential impact of a future climate change on populations of cultured bees and other pollinators in the long term, the effects of Earth's climate changes over the last millennia should be analyzed. There have been three cooling periods and three warming periods over the last 3000 years [56,57]. For example, the first global cooling took place in the Early Middle Ages; the so-called "Dark Ages" (Procopius, History of the Wars) may have resulted from a powerful volcanic eruption, probably in Indonesia, which saturated the atmosphere with dust and toxic gases. It was followed by a global warming between the 10th and the 14th centuries, and consequently a new cold period, known as the "Little Ice Age", which continued until the mid 19th century [58,59]. During these alternating periods, which had somewhat contrasting effects on agriculture and crop zoning as well as on the flora and fauna of the moderate latitude, bees did not disappear completely, even though they had undergone some population fluctuations, as can be seen from the historical documents cited above. The current warming, which has been claimed to have been around since 1850, should not have had an adverse effect on insects, including those involved in plant pollination. Climate change can impact on honey bees at different levels. It can have a direct influence on honey bee behavior and physiology [60]. It can alter the quality of the floral environment and increase or reduce colony harvesting capacity and development. It can define new honey bee distribution ranges and give rise to new competitive relationships among species and races, as well as among their parasites and pathogens [60]. Beekeepers will also be obliged to change their apiculture methods. They will favor moving their hives to new foraging areas and importing foreign races to test their value in the new environments.

Another study has investigated climate's influence on honey bee winter mortality rates across Austria [61]. The results have shown statistical correlations between monthly climate variables and winter mortality rates indicated that warmer and drier weather conditions in the preceding year were accompanied by increased winter mortality.

5.2.2. Unfavorable Weather Conditions

In addition to historical data, a number of contemporary authors [62–64] also attach great importance to unfavorable weather conditions and dramatic changes in weather as a factor in the reduction of bee colonies. Long periods of cold and rainy or hot and dry weather are associated with the sudden and unexplained disappearance of bees and the emptying of hives [28]. A probable cause is the prolonged lack of nectar and pollen (foraging), which inevitably leads to the collapse of bee colonies. The cold spring of 2013 in the USA led to a later development of hives, thus shortening the period for development of *V. destructor*. The smaller number of developed generations of the parasite, as well as the subsequent warm period suitable for active foraging time, contributed to a significantly lower loss of bee colonies in 2013 compared to the following 2014 [65,66]. In the same connection, but in the opposite direction, there also is the suggestion that chronic hive intoxication due to the treatment with thiomethoxan (a neonicotinoid), may delay the development of the larvae and endangering the existence of the bee colony [67].

Between 2006 and 2011, about 32% of hives in the USA kept dying every year due to unsuccessful wintering [3]. The same was valid for the winter of 2012–2013, when 31% of the hives perished. According to beekeepers, a symptomatic pattern similar to the "empty hive" can be observed in years when the temperature rises early in the spring, often before the snow melts. During this period of the year, bees need more water, including for breeding offspring. On the other hand, the effects of low temperatures and wind can also hinder the return to the hive, resulting in mass losses. To avoid this particular mechanism of emptying the hive, beekeepers recommend providing warmed liquids with added sugar or honey.

6. Pesticides as a Factor in the Deterioration of the Health Status of Bees and Bee Colonies

Bees are exposed to pesticides—through the chemical means used by beekeepers to control diseases and pests in or around hives. The contrasting results with clothianidin treated seeds were reported because this pesticide primarily influenced arthropod communities during the four weeks following planting, with disruptions to major natural enemy taxa, but communities showed trends toward recovery at the later corn stages. [68,69]. Rundlof, et al.'s [68] study took place in spring and reported negative effects on bees, whereas Sterk, et al.'s [69] study took place in autumn and reported no significant effects. Bees come in contact with pesticides when drifting (driven by air currents) through spray drift or dusting from an applied product, for example, in proximity to recently sprayed growing plants, when soil is treated or when treated seed is used for sowing [70,71]. Oral intoxication is also possible when visiting flowering weed vegetation recently treated with pesticide or when ingesting contaminated food and water—pollen, nectar, guttation drops, and honeydew on treated plants, etc. [72–74].

Sublethal doses of pesticide products, or even some pesticides considered to be completely safe, can lead to severe losses or endanger the existence of the bee colonies [75]. According to some observations, if during the treatment of apiaries is adjacent to the other apiaries, bees fly through the spray drift of the pesticide, and then they absorb with their bodies an odor different from the one specific to the hive [76,77]. Impacted individuals are, thus, not recognized by the bees guarding the bee colony, and, as a result of which, they are not allowed into the hive or are killed as invaders [78–80]. This circumstance further necessitates the strict application of the requirement to broadcast warnings locally before carrying out all chemical treatments—spraying, sowing of treated seeds, etc. It is obvious that the phenophase of the target culture—flowering or other—is irrelevant in this case.

6.1. Effect of Pesticides (Other than Neonicotinoids) on Bee Health

Exposure of bees to sublethal doses of pesticides occurs not only with systemic products (most commonly, aqueous solutions applied to vegetative plants) but also with non-systemics, such as pyrethroids and organophosphates, which can reduce the lifespan of individuals [75,76]. Recent studies have shown that low levels of intoxication—oral or contact, with active substances other than neonicotinoids—can weaken the immune system of exposed individuals, impair their ability to learn by monitoring and communicating with other bees, thus leading to memory loss and a change in their eating behavior and ability to distinguish flavors [72–84].

Increased levels of chlorothalonil fungicide have been detected in bee pollen in hives where high mortality has been observed. However, artificial feeding of larvae and adults with a product containing the same pollen does not cause an increase in mortality among individuals during an experiment [85]. The bees are more susceptible to most insecticides but particularly to fipronil (the most deadly to bees), most neonicotinoids and pyrethroids, and some organophosphates. According to Smith et al. [35], the greatest challenge for researchers is to extrapolate data on the effects of pesticide action on bees obtained at the individual experimentation level to the whole bee colony. Studies on their effects on bees are approved by OECD (Organisation for Economic Co-operation and Development) toxicity test guidelines. By DNA microarray analysis of transcribed products from the intestinal tract of beehives in decay, Johnson, et al. [86] found the presence of unusual RNA fragments that are thought to result from infection with one or more viruses. The authors did not find increased expression of genes related to the body's response to pesticide intoxication. For at least two decades, individual researchers have focused on exploring the mechanisms of bee detoxification and, in particular, on the role of the P450 gene, which encodes the ability of bees to metabolize toxic compounds that have fallen into the hive [87,88]. It has been suggested that there is a risk of honey bee survival if honey bees collect and drink water from water puddles in crops treated with neonicotinoid insecticides [89]. Also, honey bees and other native pollinators are threatened by cumulative exposure to these insecticides from residues in pollen, nectar, and water [89].

6.2. Neonicotinoid Insecticides

Neonicotinoids are synthetic alkaloid insecticides, analogs of natural nicotine. In treated plants, they systematically propagate acropetally (ascending xylem). They are placed in the group of neurotoxic insecticides. Neonicotinoids bind to the nicotinic receptors located in the postsynaptic membrane of neurons to cause their activation because they are agonists of the receptor [90,91]. Indeed, they are selective because they tend to bind preferentially to one of the subunits ($\alpha 4\beta 2$) that make up the receptor, which happens to be more common in insects than in vertebrates. Neonicotinoids are applied in different forms—spraying of aboveground parts of plants, treatment of seeds, and application directly to the soil in a wide range of crops. Two groups of neonicotinoids are known to date: cyano-substituted (acetamiprid and thiacloprid) and nitro-substituted (imidacloprid, thiamethoxam, clothianidin, nitenpyram, and dinotefuran). Furthermore, it has been demonstrated that the three main neonicotinoids used in agriculture the world over, namely imidacloprid, thiamethoxam, and clothianidin, pose the highest risks to bees among all other pesticides [92,93]. The difference is in the chemical formula determining the different toxicity of the two groups of active substances against bees [92,93]. For example, cyano-substituted neonicotinoids, like thiacloprid and acetamiprid, increase their toxicity to bees by 500 and 100 times in the presence of azole fungicides.

From the dawn of this millennium, a broad debate has started within the scientific community, giving rise to a number of research programs on the negative impact of neonicotinoid pesticides on pollinating insects, in particular, the managed honey bee or otherwise exploited wild bee colonies [3,38,94,95].

Neonicotinoid insecticides enter the body of honey bees, bumble bees, or other pollinators when insects feed on nectar and pollen from treated plants [96]. Krupke, et al. [97] considered different routes of contamination of beehives with neonicotinoids but focused primarily on the use of coated seeds and granules, which contaminate the soil. Yang, et al. [98] found a negative effect of sublethal doses of imidacloprid on honey bee behavior. Feeding individuals with sugar syrup containing 50 µg/L has shown to prolong the time interval between two visits to the feeding site [99,100]. The use of 1250 µg/L imidacloprid syrup for food results in significant behavioral changes, with some individuals failing to return to their normal foraging habits, while for others, the time required to return to the hive is extended significantly [99,100]. Increased mortality as a result of disorientation and inability to find the way back to the hive has been observed in bees intoxicated with thiamethoxam [101,102]. Bee colonies exposed to chronic effects of clothianidin and thiamethoxam experience significant, detrimental short and long-term impacts on colony performance and queen fate, which suggests that neonicotinoids may contribute to colony weakening in a complex manner [46]. The same authors have found that losses in individual bee colonies are determined by different levels of genetically determined resistance of bees to intoxication [46].

Some comparisons are needed with regard to the usage of neonicotinoides in different countries. In Australia, 80%–90% of beekeepers avoid agricultural fields and place their hives in forests to get the best harvest of honey. Therefore, their exposure to pesticides is minimal, unlike what occurs in the USA, Europe, Japan, or China. Australian beekeepers also know that whenever they take their hives for pollination of almonds, they lose hives, and when their bees forage on neonicotinoid-treated canola fields, they get sick and lose more hives the following winter. Moreover, Australia has been fortunate, to date, to avoid any incursion of Varroa, which presents a major threat to the health of honey bees [103]. For this reason, Australia exports bee products, bee queens, and whole hives for pollinating crops. On the contrary, imports are strictly prohibited, and quarantine measures are mandatory [104]. Because the use of neonicotinoid insecticides have proven to have particularly harmful effects on the environment (especially pollinators, and that includes bees above all). France becomes the first country in Europa that banned the use of five neonicotinoids completely.

7. Interactive and Cumulative Effects: Action of Biotic and Abiotic Stressors

With regard to the causes of death of bee colonies, opinions are most often polarized, claiming that one or another individual stress factor is the main, if not the sole, cause of the phenomenon. Recently, it has become increasingly accepted that the combined action of two or more adverse factors of different nature increases the risk of colony collapse. It has been hypothesized that the poor health status of bees is the result of individual or combined action of various factors such as stress due to poor nutrition, fasting and "monocultural" diet, abrupt meteorological changes, reduced genetic diversity in honey bee populations, etc., not excluding additional, chronic pesticide intoxication [30,39]. Too little is known about the immune response in bees at the individual and colony level. However, as social insects, bees can rely on a collective immune response to protect the colony as a whole [105]. Recently, balanced feeding of pollen and propolis has been found to be able to activate detoxifying enzymes in the individual bee [106]. It is also considered that experimental data on the effects on the protective capabilities of an individual cannot be automatically extrapolated to the entire colony in actual field experiments [35].

Recent studies have shown that interactions between pesticides and pathogens lead to deterioration in the health status of bee colonies [74,107,108]. Exposure to neonicotinoid pesticides increases the sensitivity of bees to the intestinal parasite *N. ceranae* [109]. Imidacloprid is able to synergistically increase the level of infection with Nosema spp. [73], as well as mortality [107], when both stressors are present simultaneously in the hive. Similarly, Aufauvre, et al. [74] found higher mortality from fipronil intoxication and infection with *N. ceranae* combined than when the two agents acted in isolation.

The interaction between stressors is not limited to pesticides and pathogens. Very often, the malnutrition of beehives is highlighted as a factor increasing the losses caused by bee parasites. For example, the parasite Crithidia spp. causes less mortality if wild bees have a complete food source [110]. Goulson, et al. [30] suggest that nutritional stress influences LD50 values (lethal dose, 50%) for individual bee toxic compounds. This is evidenced by the varying values of LD50 for pesticides in separate, independent studies [75].

Goulson, et al. [30], note that the individual factors that have a negative effect on bee health do not act in isolation. Obviously, all types of bees are subjected to different stress factors at the same time and with an accumulating effect over time. In doing so, each individual factor reduces the ability of bees to overcome the negative effects of the action of other stressors. The mortality of bees and bee colonies is likely to be lower if the parasite-infested hive is not further exposed to sublethal doses of toxic substances, incl. antibiotics and acaricides used in beekeeping. Moreover, the pesticides in agriculture and/or bees are not starved or subjected to a monotonous diet, often as a result of adverse weather conditions, such as prolonged drought or low temperatures. The conclusion is that complex causes require the search for complex solutions to the problem. The strategy should be aimed at reducing the general and individual stress from the action of various adverse factors by radically changing the environment in which the bees live and in which they perform their functions.

8. Some Examples for Solving the Problem of Honey Bee Population Decline

8.1. The Hindu Kush Lessons

The prospect of life without bees and other pollinators was demonstrated in the Maoxian region, Sichuan province, Southwestern China, part of the Hindu Kush Mountain, where bees, both wild and honey-bearing (European and Eastern,) disappeared more than 20 years ago [111]. In the early 1990s, local, until then self-sufficient, farmers, largely catering to their own needs, set out to create market-oriented apple and pear plantations. Both fruits are self-sterile, which requires pollinating the flowers with pollen from other, genetically distant species (varieties) of the species. With the intensification of production and the desire to market better looking fruits, the use of pesticides, and in particular, insecticides, increased. The mistaken perception that the cause of the lower pollination is insect pests attacking the flowers has led to an even more intensive use of insecticides. Natural habitats—alternative sources of nectar for bees—mainly forests and natural vegetation, have been replaced by new, industrial plantations. This effectively reduced the bee feeding period to 14 days a year. The second big problem in the region was the lack of managed honey bees and other pollinating insects. As of 1999, the problem covered neighboring Hindu Kush areas—territories of India, Pakistan, and Nepal [111,112]. The case is unique in that, in order to survive, local farmers were forced to switch to manual pollination of fruit tree flowers [111]. In India, Nepal, and Pakistan, the problem was solved with the restoration of native vegetation, providing habitats and food sources for bees, appropriate management of the natural pollination process, and training of farming colonies. Thus, bee populations were restored, and after 2011, manual pollination has rarely been practiced.

What conclusions can be drawn from the Hindu Kush experience? The main problem is, however, the use of certain pesticides in agriculture, even as recommended on the labels [113]. Another problem is the approval of certain compounds and formulations for use in some crops because they pose more risks to bees and the environment than those estimated by the regulatory authorities. The latter is due to authorities using insufficient or inappropriate information and out of date methods [114]. Improper use of pesticides in agriculture is part of the problem, but it is not the only cause of death for honey and other bee species. Refusing to use pesticides, even if possible and as a sole measure, would not be a solution to the problem. Restoring and preserving the natural habitats of pollinators today can ensure the diversity of our table tomorrow [115].

8.2. The Experience of North America

With the decrease in the number of bee colonies in the United States by more than 50% as a result of the introduction of parasitic mites, heavy pesticide use, and industrialized plantations, crossbreeds between the European, *A. mellifera* and the African honey bee, *Apis mellifera scutellata* Lepeletier have been introduced [116]. New, hybrid forms show increased resistance to parasites, infectious diseases, and some pesticides [117,118]. They also find successful application in the agricultural practices of Latin American countries, such as Brazil, where they have replaced the European bee and have even been rated as better pollinators than the European one [119]. The problems associated with the use of these: the so-called "Africanized" bees or "killer bees" include smaller numbers of workers in one colony, shorter flight and perimeter of feeding around the colonies, and high mortality when moving to a new foraging site. In addition, Africanized bees are extremely aggressive. They are now considered inapplicable to the USA because of the mobile nature of bee pollination used there for many crops [117]. The measures taken in the country are aimed at the use of less intensive forms of agriculture related to reducing the dependence of agriculture on the use of pesticides. The conservation of weeds and other vegetation around arable land where bees would find additional foraging for a longer period is promoted. Research in this regard shows that structural changes in the landscape distribution of crops with the conservation and/or introduction of wild plant species increase the species diversity of bees and pollinating butterflies [120].

9. Concluding Remarks

The recent concerns in some regions over an increase in colony losses have prompted investment in more coordinated monitoring of bees and research into how pests and diseases, bee diversity, beekeeping practices, and their foraging environment is affecting bee vitality. The global picture shows that regions with established honey bee parasitic Varroa mite populations (*V. destructor*) have consistently higher colony losses. The active role of *V. destructor* as a vector of bee viruses is emerging as a significant factor in the losses of honey bee colonies seen globally. In addition, land management and environmental conditions affect the availability and quality of food sources and also affect conditions in the hive, and effective management of bee colonies under changing situations is dependent on beekeeping practices and bee selection/breeding.

The potential pesticide risk maps could aid in the prevention of honeybee colony losses. The maps could help identify regions with relatively high pesticide pressure in a species-specific manner, enabling conservation actions on a local scale. This could result in a lower local pesticide pressure for the bee species in question while minimizing economic damages since enforced pesticide regulations could be tailored to local high-risk areas. Such actions could help restore threatened bee species and lower honey bee colony losses, which would benefit pollinator dependent farmers and plant species by the increase of pollination services.

Author Contributions: All authors contributed equally to writing this review.

Funding: This work was funded by the National Scientific Fund of the Bulgarian Ministry of Education and Science (grant numbers 06/10 17/12/2016). The funders had no role in study design, data collection and analysis, decision to publish, or preparation of the manuscript.

Conflicts of Interest: The authors declare no conflict of interest.

References

1. Crane, E. Recent research on the world history of beekeeping. *Bee World* **1999**, *80*, 174–186. [CrossRef]
2. Roffet-Salque, M.; Regert, M.; Evershed, R.P.; Outram, A.K.; Cramp, L.J.; Decavallas, O.; Dunne, J.; Gerbault, P.; Mileto, S.; Mirabaud, S.; et al. Widespread exploitation of the honeybee by early Neolithic famers. *Nature* **2015**, *527*, 226–230. [CrossRef] [PubMed]
3. Bee Health: The Role of Pesticides; Reports for Congress. Available online: https://www.fas.org/sgp/crs/misc/R42855.pdf (accessed on 11 December 2012).
4. Potts, S.G.; Biesmeijer, J.C.; Kremen, C.; Neumann, P.; Schweiger, O.; Kunin, W.E. Global pollinator declines: Trends, impacts and drivers. *Trends Ecol. Evol.* **2010**, *25*, 345–353. [CrossRef] [PubMed]
5. Gill, R.J.; Ramos-Rodriguez, O.; Raine, N.E. Combined pesticide exposure severely affects individual- and colony-level traits in bees. *Nature* **2012**, *491*, 105–108. [CrossRef] [PubMed]
6. Ollerton, J.; Winfree, R.; Tarrant, S. How many flowering plants are pollinated by animals? *Oikos* **2011**, *120*, 321–326. [CrossRef]
7. Geslin, B.; Gauzens, B.; Baude, M.; Dajoz, I.; Fontaine, C.; Henry, M.; Ropars, L.; Rollin, O.; Thébault, E.; Vereecken, N.J. Massively introduced managed species and their consequences for plant–pollinator interactions. *Adv. Ecol. Res.* **2017**, *57*, 147–199.

8. Hermansen, T.D.; Britton, D.R.; Ayre, D.J.; Minchinton, T.E. Identifying the real pollinators? Exotic honey bees are the dominant flower visitors and only effective pollinators of Avicennia marina in Australian temperate mangroves. *Estuar. Coasts* **2014**, *37*, 621–635. [CrossRef]

9. Thompson, C.E.; Biesmeijer, J.C.; Allnutt, T.R.; Pietravalle, S.; Budge, G.E. Parasite pressures on feral honey bees (*Apis mellifera* sp.). *PLoS ONE* **2014**, *9*, e105164. [CrossRef]

10. Giannini, T.C.; Garibaldi, L.A.; Acosta, A.L.; Silva, J.S.; Maia, K.P.; Saraiva, A.M.; Guimaraes, P.R.; Kleinert, A.M.P. Native and non-native supergeneralist bee species have different effects on plant-bee networks. *PLoS ONE* **2015**, *10*, e0137198. [CrossRef]

11. Abe, T.; Wada, K.; Kato, Y.; Makino, S.; Okochi, I. Alien pollinator promotes invasive mutualism in an insular pollination system. *Biol. Invasions* **2011**, *13*, 957–967. [CrossRef]

12. Gross, C.L.; Mackay, D. Honeybees reduce fitness in the pioneer shrub *Melastoma affine* (Melastomataceae). *Biol. Conserv.* **1998**, *86*, 169–178. [CrossRef]

13. Wratten, S.D.; Gillespie, M.; Decourtye, A.; Mader, E.; Desneux, N. Pollinator habitat enhancement: Benefits to other ecosystem services. *Agric. Ecosyst. Environ.* **2012**, *159*, 112–122. [CrossRef]

14. Rader, R.; Bartomeus, I.; Garibaldi, L.A.; Garratt, M.P.; Howlett, B.G.; Winfree, R.; Cunningham, S.A.; Mayfield, M.M.; Arthur, A.D.; Andersson, G.K.; et al. Non-bee insects are important contributors to global crop pollination. *Proc. Natl. Acad. Sci. USA* **2016**, *113*, 146–151. [CrossRef] [PubMed]

15. Minteer, B.A.; Collins, J.P. Move it or lose it? The ecological ethics of relocating species under climate change. *Ecol. Appl.* **2010**, *20*, 1801–1804. [CrossRef] [PubMed]

16. UNEP Emerging Issues: Global Honey Bee Colony Disorder and Other Threats to Insect Pollinators. Available online: http://www.unep.org/dewa/Portals/67/pdf/Global_Bee_Colony_Disorder_and_Threats_insect_pollinators.pdf (accessed on 10 March 2011).

17. Rehan, S.M.; Leys, R.; Schwarz, M.P. First Evidence for a Massive Extinction Event Affecting Bees Close to the K-T Boundary. *PLoS ONE* **2013**, *8*, e76683. [CrossRef]

18. Fleming, G. *Animal Plagues: Their History, Nature, and Prevention*; Chapman and Hall: London, UK, 1871; p. 548.

19. Pellett, F.C. *History of American Beekeeping*; Collegiate Press: Ames, IA, USA, 1938; p. 303.

20. Critchlow, B.P. Gleanings in Bee Culture. *Bee Cult. Mag.* **1904**, *32*, 692.

21. Finley, J.; Camazine, S.; Frazier, M. The epidemic of honey bee colony losses during the 1995–1996 season. *Am. Bee J.* **1996**, *136*, 805–808.

22. Bailey, L. The 'Isle of Wight Disease': The Origin and Significance of the Myth. *Bee World* **1964**, *45*, 32–37. [CrossRef]

23. Bailey, L. *The Isle of Wight Disease*; Central Association of Bee-Keepers: Poole, UK, 2002; p. 11.

24. Aizen, M.A.; Harder, L.D. The global stock of domesticated honey bees is growing slower than agricultural demand for pollination. *Curr. Biol.* **2009**, *19*, 915–918. [CrossRef]

25. Sumner, D.A.; Boriss, H. Bee-conomics and the leap in pollination fees. *J. Agric. Resour. Econ.* **2006**, *9*, 9–11.

26. Daberkow, S.; Korb, P.; Hoff, F. Structure of the US beekeeping industry: 1982–2002. *J. Econ. Entomol.* **2009**, *102*, 868–886. [CrossRef] [PubMed]

27. vanEngelsdorp, D.; Caron, D.; Hayes, J.; Underwood, R.; Henson, M.; Rennich, K.; Spleen, A.; Andree, M.; Snyder, R.; Lee, K.; et al. A national survey of managed honey bee 2010-2011 winter colony losses in the USA: Results from the Bee Informed Partnership. *J. Apic. Res.* **2012**, *51*, 115–124. [CrossRef]

28. vanEngelsdorp, D.; Evans, J.D.; Saegerman, C.; Mullin, C.; Haubruge, E.; Nguyen, B.K.; Frazier, M.; Frazier, J.; Cox-Foster, D.; Chen, Y.P.; et al. Colony collapse disorder: A descriptive study. *PLoS ONE* **2009**, *4*, e6481. [CrossRef] [PubMed]

29. vanEngelsdorp, D.; Hayes, J.J.; Underwood, R.; Pettis, J.S. A survey of honey bee colony losses in the United States, fall 2008 to spring 2009. *J. Apic. Res.* **2010**, *49*, 7–14. [CrossRef]

30. Goulson, D.; Nicholls, E.; Botías, C.; Rotheray, E.L. Bee Declines Driven by Combined Stress from Parasites, Pesticides and Lack of Flowers. *Science* **2015**, *347*, 1255957. [CrossRef]

31. Le Conte, Y.; Ellis, M.; Ritter, W. Varroa mites and honey bee health: Can Varroa explain part of the colony losses? *Apidologie* **2010**, *41*, 353–363. [CrossRef]

32. Fries, I.; Martín-Hernández, R.; Meana, A.; García-Palencia, P.; Higes, M. Natural infections of *Nosema ceranae* in European honey bees. *J. Apic. Res.* **2006**, *45*, 230–233. [CrossRef]

33. Ellis, J.D.; Evans, J.D.; Pettis, J. Colony losses, managed colony population decline, and Colony Collapse Disorder in the United States. *J. Apic. Res.* **2010**, *49*, 134–136. [CrossRef]

34. van der Zee, R.; Pisa, L.; Andonov, S.; Brodschneider, R.; Charriere, J.D.; Chlebo, R.; Coffey, M.F.; Crailsheim, K.; Dahle, B.; Gajda, A.; et al. Managed honey bee colony losses in Canada, China, Europe, Israel and Turkey, for the winters of 2008–9 and 2009–10. *J. Apic. Res.* **2012**, *51*, 91–114. [CrossRef]

35. Smith, K.M.; Loh, E.H.; Rostal, M.K.; Zambrana-Torrelio, C.M.; Mendiola, L.; Daszak, P. Pathogens, Pests, and Economics: Drivers of Honey Bee Colony Declines and Losses. *EcoHealth* **2014**, *10*, 434–445. [CrossRef]

36. Brodschneider, R.; Brus, J.; Danihlík, J. Comparison of apiculture and winter mortality of honey bee colonies (*Apis mellifera*) in Austria and Czechia. *Agric. Ecosyst. Environ.* **2019**, *274*, 24–32. [CrossRef]

37. Capri, E.; Marchis, A. Bee Health in Europe: Facts and Figures 2013. In *Compendium of the Latest Information on Bee Health in Europe*; OPERA Research Centre, Università Cattolica del Sacro Cuore: Milano, Italy, 2013; p. 64.

38. Potts, S.; Biesmeijer, K.; Bommarco, R.; Breeze, T.; Carvalheiro, L.; Franzen, M.; Gonzalez-Varo, J.P.; Holzschuh, A.; Kleijn, D.; Klein, A.-M.; et al. Status and trends of European pollinators. In *Key Findings of the STEP Project*; Pensoft Publishers: Sofia, Bulgaria, 2015; p. 72.

39. Bee Health: The Role of Pesticides; Congressional Research Service (CRS), Reports for Congress. Available online: https://www.fas.org/sgp/crs/misc/R43900.pdf (accessed on 17 February 2015).

40. Anderson, D.L.; Trueman, J.W.H. *Varroa jacobsoni* (Acari: Varroidae) is more than one species. *Exp. Appl. Acarol.* **2000**, *24*, 165–189. [CrossRef] [PubMed]

41. Sanford, M.T.; Demark, H.A.; Cromroy, H.L.; Cutts, L. Featured Creatures: Varroa mite. USA: University of Florida Institute of Food and Agricultural Science. 2007. Available online: http://creatures.ifas.ufl.edu/misc/bees/Varroa_mite.htm (accessed on 5 November 2019).

42. Rosenkranz, P.; Aumeier, P.; Ziegelmann, B. Biology and control of *Varroa destructor*. *J. Invertebr. Pathol.* **2010**, *103*, S96–S119. [CrossRef] [PubMed]

43. Neumann, P.; Carreck, N.L. Honey bee colony losses. *J. Apic. Res.* **2010**, *49*, 1–6. [CrossRef]

44. Aston, D. Honey bee winter loss survey for England, 2007-8. *J. Apic. Res.* **2010**, *49*, 111–112. [CrossRef]

45. Gajger, I.T.; Tomljanovic, Z.; Petrinec, Z. Monitoring health status of Croatian honey bee colonies and possible reasons for winter losses. *J. Apic. Res.* **2010**, *49*, 107–108. [CrossRef]

46. Genersch, E.; Aubert, M. Emerging and re-emerging viruses of the honey bee (*Apis mellifera* L.). *Vet. Res.* **2010**, *41*, 54. [CrossRef]

47. Genersch, E.; von der Ohe, W.; Kaatz, H.; Schroeder, A.; Otten, C.; Buchler, R.; Berg, S.; Ritter, W.; Muhlen, W.; Gisder, S.; et al. The German bee monitoring project: A long term study to understand periodically high winter losses of honey bee colonies. *Apidologie* **2010**, *41*, 332–352. [CrossRef]

48. Martin, S.J.; Highfield, A.C.; Brettell, L.; Villalobos, E.M.; Budge, G.E.; Powell, M.; Nikaido, S.; Schroeder, D.C. Global honey bee viral landscape altered by a parasitic mite. *Science* **2012**, *336*, 1304–1306. [CrossRef]

49. Paxton, R.J. Does infection by Nosema ceranae cause "Colony Collapse Disorder" in honey bees (*Apis mellifera*)? *J. Apic. Res.* **2010**, *49*, 80–84. [CrossRef]

50. Kaplan, J.K. Colony Collapse Disorder: An Incomplete Puzzle, Agricultural Research (USDA publication). 2012. Available online: http://www.ars.usda.gov/is/AR/archive/jul12/colony0712.htm (accessed on 5 November 2019).

51. Higes, M.; Garcia-Palencia, P.; Martin-Hernandez, R.; Meana, A. Experimental infection of *Apis mellifera* honeybees with *Nosema ceranae* (Microsporidia). *J. Invertebr. Pathol.* **2007**, *94*, 211–217. [CrossRef] [PubMed]

52. Ellis, J.D.; Spiewok, S.; Delaplane, K.S.; Buchholz, S.; Neumann, P.; Tedders, W.L. Susceptibility of *Aethina tumida* (Coleoptera: Nitidulidae) larvae and pupae to entomopathogenic nematodes. *J. Econ. Entomol.* **2010**, *103*, 1–9. [CrossRef] [PubMed]

53. Kwadha, C.A.; Ongamo, G.O.; Ndegwa, P.N.; Raina, S.K.; Fombong, A.T. The Biology and Control of the Greater Wax Moth, Galleria mellonella. *Insects* **2017**, *8*, 61. [CrossRef] [PubMed]

54. vanEngelsdorp, D.; Hayes, J.; Underwood, R.M.; Caron, D.; Pettis, J. A survey of managed honey bee colony losses in the USA, fall 2009 to winter 2010. *J. Apic. Res.* **2011**, *50*, 1–10. [CrossRef]

55. Cox-Foster, D.L.; Conlan, S.; Holmes, E.C.; Palacios, G.; Evans, J.D.; Moran, N.A.; Quan, P.L.; Briese, T.; Hornig, M.; Geiser, D.M.; et al. A metagenomic survey of microbes in honey bee colony collapse disorder. *Science* **2007**, *318*, 283–287. [CrossRef]

56. Broecker, W.S. Was the medieval warm period global? *Science* **2001**, *291*, 1497–1499. [CrossRef]

57. Andrew, J.T.; Norman, D.Y.; Keller, B.; Girard, R.; Heneberry, J.; Gunn, J.M.; Hamilton, D.P.; Taylor, P.A. Cooling lakes while the world warms: Effects of forest regrowth and increased dissolved organic matter on the thermal regime of a temperate, urban lake. *Limnol. Oceanogr.* **2008**, *53*, 404–410. [CrossRef]

58. Matthews, J.A.; Briffa, K.R. The 'Little Ice Age': Re-evaluation of an evolving concept. *Geogr. Ann. A* **2005**, *87*, 17–36. [CrossRef]

59. Mann, M.E.; Zhang, Z.; Rutherford, S.; Bradley, R.S.; Hughes, M.K.; Shindell, D.; Ammann, C.; Faluvegi, G.; Ni, F. Global signatures and dynamical origins of the Little Ice Age and Medieval Climate Anomaly. *Science* **2009**, *326*, 1256–1260. [CrossRef]

60. Le Conte, Y.; Navajas, M. Climate change: Impact on honey bee populations and diseases. *Rev. Sci. Tech. OIE* **2008**, *27*, 499–510.

61. Switanek, M.; Crailsheim, K.; Truhetz, H.; Brodschneider, R. Modelling seasonal effects of temperature and precipitation on honey bee winter mortality in a temperate climate. *Sci. Total. Environ.* **2017**, *579*, 1581–1587. [CrossRef] [PubMed]

62. Bartomeus, I.; Ascher, J.S.; Wagner, D.; Danforth, B.N.; Colla, S.; Kornbluth, S.; Winfree, R. Climate-associated phenological advances in bee pollinators and bee-pollinated plants. *Proc. Natl. Acad. Sci. USA* **2011**, *108*, 20645–20649. [CrossRef] [PubMed]

63. Guler, Y.; Dikmen, F. Potential bee pollinators of sweet cherry in inclement weather conditions. *J. Entomol. Res. Soc.* **2013**, *15*, 9–19.

64. Zagareanu, A.; Mardari, T.; Modvala, S. Stimulation of resistance of bee families during wintering. *J. Anim. Sci. Biotechnol.* **2013**, *46*, 268–271.

65. Brodschneider, R.; Gray, A.; Adjlane, N.; Ballis, A.; Brusbardis, V.; Charrière, J.-D.; Chlebo, R.; Coffey, M.F.; Dahle, B.; de Graaf, D.C.; et al. Multi-country loss rates of honey bee colonies during winter 2016/2017 from the COLOSS survey. *J. Apic. Res.* **2018**, *57*, 452–457. [CrossRef]

66. Gray, A.; Brodschneider, R.; Adjlane, N.; Ballis, A.; Brusbardis, V.; Charrière, J.-D.; Chlebo, R.; Coffey, M.F.; Cornelissen, B.; da Costa, C.A.; et al. Loss rates of honey bee colonies during winter 2017/18 in 36 countries participating in the COLOSS survey, including effects of forage sources. *J. Apic. Res.* **2019**, *58*, 479–485. [CrossRef]

67. Alburaki, M.; Boutin, S.; Mercier, P.L.; Loublier, Y.; Chagnon, M.; Derome, N. Neonicotinoid-coated *Zea mays* seeds indirectly affect honeybee performance and pathogen susceptibility in field trials. *PLoS ONE* **2015**, *10*, e0125790. [CrossRef]

68. Rundlöf, M.; Andersson, G.K.; Bommarco, R.; Fries, I.; Hederström, V.; Herbertsson, L.; Jonsson, O.; Klatt, B.K.; Pedersen, T.R.; Yourstone, J.; et al. Seed coating with a neonicotinoid insecticide negatively affects wild bees. *Nature* **2015**, *521*, 77. [CrossRef]

69. Sterk, G.; Peters, B.; Gao, Z.; Zumkier, U. Large-scale monitoring of effects of clothianidin-dressed OSR seeds on pollinating insects in Northern Germany: Effects on large earth bumble bees (Bombus terrestris). *Ecotoxicology* **2016**, *25*, 1666–1678. [CrossRef]

70. Nuyttens, D.; Devarrewaere, W.; Verboven, P.; Foque, D. Pesticide-laden dust emission and drift from treated seeds during seed drilling: A review. *Pest. Manag. Sci.* **2013**, *69*, 564–575. [CrossRef]

71. Sanchez-Bayo, F.; Goka, K. Pesticide residues and bees–a risk assessment. *PLoS ONE* **2014**, *9*, e94482. [CrossRef] [PubMed]

72. Alaux, C.; Ducloz, F.; Crauser, D.; Le Conte, Y. Diet effects on honeybee immunocompetence. *Biol. Lett.* **2010**, *6*, 562–565. [CrossRef] [PubMed]

73. Aufauvre, J.; Biron, D.G.; Vidau, C.; Fontbonne, R.; Roudel, M.; Diogon, M.; Vigues, B.; Belzunces, L.P.; Delbac, F.; Blot, N. Parasite-insecticide interactions: A case study of Nosema ceranae and fipronil synergy on honeybee. *Sci. Rep.* **2012**, *2*, 326. [CrossRef] [PubMed]

74. Pisa, L.W.; Amaral-Rogers, V.; Belzunces, L.P.; Bonmatin, J.M.; Downs, C.A.; Goulson, D.; Kreutzweiser, D.P.; Krupke, C.; Liess, M.; McField, M. Effects of neonicotinoids and fipronil on non-target invertebrates. *Environ. Sci. Pollut. Res. Int.* **2015**, *22*, 68–102. [CrossRef]

75. Meikle, W.G.; Corby-Harris, V.; Carroll, M.J.; Weiss, M.; Snyder, L.A.; Meador, C.A.D.; Beren, E.; Brown, N. Exposure to sublethal concentrations of methoxyfenozide disrupts honey bee colony activity and thermoregulation. *PLoS ONE* **2019**, *14*, e0204635. [CrossRef]

76. Zawislak, J.; Adamczyk, J.; Johnson, D.R.; Lorenz, G.; Black, J.; Hornsby, Q.; Stewart, S.D.; Joshi, N. Comprehensive Survey of Area-Wide Agricultural Pesticide Use in Southern United States Row Crops and Potential Impact on Honey Bee Colonies. *Insects* **2019**, *10*, 280. [CrossRef]

77. Cook, S.C. Compound and Dose-Dependent Effects of Two Neonicotinoid Pesticides on Honey Bee (Apis mellifera) Metabolic Physiology. *Insects* **2019**, *10*, 18. [CrossRef]

78. Li, Z.; Li, M.; He, J.; Zhao, X.; Chaimanee, V.; Huang, W.F.; Nie, H.; Zhao, Y.; Su, S. Differential physiological effects of neonicotinoid insecticides on honey bees: A comparison between *Apis mellifera* and *Apis cerana*. *Pestic. Biochem. Physiol* **2017**, *140*, 1–8. [CrossRef]

79. Nasuti, C.; Fattoretti, P.; Carloni, M.; Fedeli, D.; Ubaldi, M.; Ciccocioppo, R.; Gabbianelli, R. Neonatal exposure to permethrin pesticide causes lifelong fear and spatial learning deficits and alters hippocampal morphology of synapses. *J. Neurodev. Dis.* **2014**, *6*, 7. [CrossRef]

80. Chrustek, A.; Hołyńska-Iwan, I.; Dziembowska, I.; Bogusiewicz, J.; Wróblewski, M.; Cwynar, A.; Olszewska-Słonina, D. Current Research on the Safety of Pyrethroids Used as Insecticides. *Medicine* **2018**, *54*, 61. [CrossRef]

81. Azpiazu, C.; Bosch, J.; Viñuela, E.; Medrzycki, P.; Teper, D.; Sgolastra, F. Chronic oral exposure to field-realistic pesticide combinations via pollen and nectar: Effects on feeding and thermal performance in a solitary bee. *Sci. Rep.* **2019**, *9*, 1–11. [CrossRef] [PubMed]

82. Gradish, A.E.; Van Der Steen, J.; Scott-Dupree, C.D.; Cabrera, A.R.; Cutler, G.C.; Goulson, D.; Klein, O.; Lehmann, D.M.; Lückmann, J.; O'Neill, B. Comparison of pesticide exposure in honey bees (Hymenoptera: Apidae) and bumble bees (Hymenoptera: Apidae): Implications for risk assessments. *Environ. Entomol.* **2018**, *48*, 12–21. [CrossRef] [PubMed]

83. Williamson, S.M.; Wright, G.A. Exposure to multiple cholinergic pesticides impairs olfactory learning and memory in honeybees. *J. Exp. Biol.* **2013**, *216*, 1799–1807. [CrossRef] [PubMed]

84. vanEngelsdorp, D.; Evans, J.D.; Donovall, L.; Mullin, C.; Frazier, M.; Frazier, J.; Tarpy, D.R.; Hayes, J.; Pettis, J.S. Entombed pollen: A new condition in honey bee colonies associated with increased risk of colony mortality. *J. Invertebr. Pathol.* **2009**, *101*, 147–149. [CrossRef] [PubMed]

85. Johnson, R.M. Managed pollinator CAP Coordinated Agricultural Project: A national research and extension initiative to reverse pollinator decline when varroacides interact. *Am. Bee J.* **2009**, *149*, 1157–1159.

86. Johnson, R.M.; Dahlgren, L.; Siegfried, B.D.; Ellis, M.D. Acaricide, fungicide and drug interactions in honey bees (*Apis mellifera*). *PLoS ONE* **2012**, *8*, e54092. [CrossRef]

87. Johnson, R.M. Honey Bee Toxicology. *Ann. Rev. Entomol* **2015**, *60*, 22.1–22.20. [CrossRef]

88. Samson-Robert, O.; Labrie, G.; Chagnon, M.; Fournier, V. Neonicotinoid-Contaminated Puddles of Water Represent a Risk of Intoxication for Honey Bees. *PLoS ONE* **2014**, *9*, e108443. [CrossRef]

89. Nauen, R.; Ebbinghaus-Kintscher, U.; Salgado, V.L.; Kaussmann, M. Thiamethoxam is a neonicotinoid precursor converted to clothianidin in insects and plants. *Pestic. Biochem. Phys.* **2003**, *76*, 55–69. [CrossRef]

90. Matsuda, K.; Buckingham, S.D.; Kleier, D.; Rauh, J.J.; Grauso, M.; Sattelle, D.B. Neonicotinoids: Insecticides acting on insect nicotinic acetylcholine receptors. *Trends Pharmacol. Sci.* **2001**, *22*, 573–580. [CrossRef]

91. Iwasa, T.; Motoyama, N.; Ambrose, J.T.; Roe, R.M. Mechanism for the differential toxicity of neonicotinoid insecticides in the honey bee, *Apis mellifera*. *J. Crop. Protection.* **2004**, *23*, 371–378. [CrossRef]

92. Girolami, V.; Mazzon, L.; Squartini, A.; Mori, N.; Marzaro, M.; Di Bernardo, A.; Greatti, M.; Giorio, C.; Tapparo, A. Translocation of neonicotinoid insecticides from coated seeds to seedling guttation drops: A novel way of intoxication for bees. *J. Econ. Entomol.* **2009**, *102*, 1808–1815. [CrossRef] [PubMed]

93. Faucon, J.P.; Aurieres, C.; Drajnudel, P.; Ribiere, M.; Martel, A.C.; Zeggane, S.; Chauzat, M.P.; Aubert, M.F.A. Experimental study on the toxicity of imidacloprid given in syrup to honey bees (*Apis mellifera*). *Pest. Manag. Sci.* **2005**, *61*, 111–125. [CrossRef] [PubMed]

94. Blacquiere, T.; Smagghe, G.; van Gestel, C.; Mommaerts, V. Neonicotinoids in bees: A review on concentrations, side-effects and risk assessment. *Ecotoxicology* **2012**, *2*, 973–992. [CrossRef] [PubMed]

95. Cresswell, J.E.; Page, C.J.; Uygun, M.B.; Holmbergh, M.; Li, Y.; Wheeler, J.G.; Laycock, I.; Pook, C.J.; de Ibarra, N.H.; Smirnoff, N.; et al. Differential sensitivity of honey bees and bumble bees to a dietary insecticide (imidacloprid). *Zoology* **2012**, *115*, 365–371. [CrossRef] [PubMed]

96. Krupke, C.H.; Hunt, G.J.; Eitzer, B.D.; Andino, G.; Given, K. Multiple routes of pesticide exposure for honey bees living near agricultural fields. *PLoS ONE* **2012**, *7*, e29268. [CrossRef]

97. Yang, E.C. Abnormal foraging be¬havior induced by sublethal dosage of imidacloprid in the honey bee (Hymenoptera: Apidae). *J. Econ. Entomol.* **2008**, *101*, 1743–1748. [CrossRef]

98. Chan, D.S.; Prosser, R.S.; Rodríguez-Gil, J.L.; Raine, N.E. Assessment of risk to hoary squash bees (*Peponapis pruinosa*) and other ground-nesting bees from systemic insecticides in agricultural soil. *Sci Rep.* **2019**, *9*, 11870. [CrossRef]

99. Tosi, S.; Nieh, J.C. Lethal and sublethal synergistic effects of a new systemic pesticide, flupyradifurone (Sivanto®), on honeybees. *Proc. R. Soc. B* **2019**, *286*, 20190433. [CrossRef]

100. Guez, D. A common pesticide decreases foraging success and survival in honey bees: Questioning the ecological relevance. *Front. Physiol.* **2013**, *4*, 1–3. [CrossRef]

101. Henry, M.; Beguin, M.; Requier, F.; Rollin, O.; Odoux, J.-F.; Aupinel, P.; Aptel, J.; Sylvie Tchamitchian, S.; Decourtye, A. A common pesticide decreases foraging success and survival in honey bees. *Science* **2012**, *336*, 348–350. [CrossRef] [PubMed]

102. Sandrock, C.; Tanadini, M.; Tanadini, L.G.; Fauser-Misslin, A.; Potts, S.G.; Neumann, P. Impact of chronic neonicotinoid exposure on honeybee colony performance and queen supersedure. *PLoS ONE* **2014**, *9*, e103592. [CrossRef] [PubMed]

103. Tozzi, J. *Memorandum: Is Varroa destructor or Neonicotinoid Pesticides Responsible for Bee Health Decline?* Center for Regulatory Effectiveness: Washington, DC, USA, 2014; p. 20.

104. Watson, M.; McCarthy, M. Asian Honey Bee Swarm Carrying Varroa Mite Destroyed after Discovery in Brisbane Shipment. 2015. Available online: http://www.abc.net.au/news/2015-05-13/asian-honey-bee-swarm-destroyed-nest-found-in-malaysia-shipment/6465672 (accessed on 13 May 2015).

105. Evans, J.D.; Spivak, M. Socialized medicine: Individual and communal disease barriers in honey bees. *J. Invertebr. Pathol.* **2010**, *103*, S62–S72. [CrossRef] [PubMed]

106. Mao, W.; Shuler, M.A.; Berenbaum, M.R. Honey constituents up-regulate detoxification and immunity genes in the western honey bee *Apis mellifera*. *Proc. Natl. Acad. Sci. USA* **2013**, *110*, 8763–8764. [CrossRef]

107. Pettis, J.S.; vanEngelsdorp, D.; Johnson, J.; Dively, G. Pesticide exposure in honey bees results in increased levels of the gut pathogen Nosema. *Naturwissenschaften* **2012**, *99*, 153–158. [CrossRef]

108. Baron, G.L.; Raine, N.E.; Brown, M.J.F. Impact of chronic exposure to a pyrethroid pesticide on bumblebees and interactions with a trypanosome parasite. *J. Appl. Ecol.* **2014**, *51*, 460–469. [CrossRef]

109. Wu, J.Y.; Smart, M.D.; Anelli, C.M.; Sheppard, W.S. Honey bees (Apis mellifera) reared in brood combs containing high levels of pesticide residues exhibit increased susceptibility to Nosema (Microsporidia) infection. *J. Invertebr. Pathol.* **2012**, *109*, 326–329. [CrossRef]

110. Brown, M.J.F.; Loosli, R.; Schmid-Hempel, P. Condition-dependent expression of virulence in a trypanosome infecting bumblebees? *Oikos* **2000**, *91*, 421–427. [CrossRef]

111. Partap, U.; Ya, T. The human pollinators of fruit crops in Maoxian county, Sichuan, China. *Mt. Res. Dev.* **2012**, *32*, 176–186. [CrossRef]

112. Partap, U. Cach crop farming in the Himalayas: The importance of the pollinator management and managed pollination. In *FAO, Biodiversity and the Ecosystem Approach in Agriculture, Forestry and Fisheries*; FAO Corporate Document Repository: Rome, Italy, 2003; Available online: http://www.fao.org/docrep/005/y4586e/y4586e11.htm#P0_0 (accessed on 5 November 2019).

113. Castilhos, D.; Dombroski, J.L.D.; Bergamo, G.C.; Gramacho, K.P.; Gonçalves, L.S. Neonicotinoids and fipronil concentrations in honeybees associated with pesticide use in Brazilian agricultural areas. *Apidologie* **2019**, *50*, 657–668. [CrossRef]

114. Brühl, C.A.; Zaller, J.G. Biodiversity decline as a consequence of an inadequate environmental risk assessment of pesticides. *Front. Environ. Sci.* **2019**, *7*, 177. [CrossRef]

115. Burkle, L.A.; Marlin, J.C.; Knight, T.M. Plant-pollinator interactions over 120 years: Loss of species, co-occurrence, and function. *Science* **2013**, *339*, 1611–1615. [CrossRef] [PubMed]

116. Ghazoul, J. Buzziness as usual? Questioning the global pollination crisis. *Trends Ecol. Evol.* **2005**, *20*, 367–373. [CrossRef] [PubMed]

117. Schneider, S.S.; DeGrandi-Hoffman, G.; Smith, D.R. The African Honey Bee: Factors Contributing to a Successful Biological Invasion. *Annu. Rev. Entomol.* **2004**, *49*, 351–376. [CrossRef] [PubMed]

118. Moore, P.; Wilson, M.; Skinner, J. Africanized Bees: Better Understanding, Better Prepared. *eXtention* **2015**, 1–12.

119. Basualdo, M.; Bedascarrasbure, E.; Jong, D. Africanized honey bees (Hymenoptera: Apidae) have a greater fidelity to sunflowers than European bees. *J. Econ. Entomol.* **2000**, *93*, 304–307. [CrossRef]
120. Kremen, C.; Williams, N.M.; Bugg, R.L.; Fay, J.P.; Thorp, R.W. The area requirements of an ecosystem service: Crop pollination by native bee communities in California. *Ecol. Lett.* **2004**, *7*, 1109–1119. [CrossRef]

MDPI

St. Alban-Anlage 66

4052 Basel

Switzerland

Tel. +41 61 683 77 34

Fax +41 61 302 89 18

www.mdpi.com

Diversity Editorial Office

E-mail: diversity@mdpi.com

www.mdpi.com/journal/diversity

www.ingramcontent.com/pod-product-compliance
Lightning Source LLC
LaVergne TN
LVHW070926170726
843515LV00005B/883